Evolutionary Dynamics of Complex Communications Networks

OTHER COMMUNICATIONS BOOKS FROM AUERBACH

Evolutionary Dynamics of Complex Communications Networks

Vasileios Karyotis
Eleni Stai
Symeon Papavassiliou

CRC Press
Taylor & Francis Group
Boca Raton London New York

CRC Press is an imprint of the
Taylor & Francis Group, an **informa** business

CRC Press
Taylor & Francis Group
6000 Broken Sound Parkway NW, Suite 300
Boca Raton, FL 33487-2742

First issued in paperback 2017

ISBN-13: 978-1-4665-1840-7 (hbk)
ISBN-13: 978-1-138-03408-2 (pbk)

Library of Congress Cataloging-in-Publication Data

Karyotis, Vasileios.
  Evolutionary dynamics of complex communications networks / Vasileios Karyotis, Eleni Stai, Symeon Papavassiliou.
      pages cm
  Includes bibliographical references and index.
  ISBN 978-1-4665-1840-7 (hardback)
  1. Telecommunication systems. 2. Communication--Network analysis. I. Stai, Eleni. II. Papavassiliou, Symeon. III. Title.

TK5101.K257 2013
621.382'1--dc23   2013027682

Visit the Taylor & Francis Web site at
http://www.taylorandfrancis.com

and the CRC Press Web site at
http://www.crcpress.com

Contents

Preface   ix

1 Introduction   1
   1.1 Approach and Objectives   2
   1.2 Fundamentals of Complex Networks   5
        1.2.1 Complex Networks Fundamentals   10
        1.2.2 Complex Network Taxonomy and Examples   12
   1.3 Network Science   18
        1.3.1 Content and Promise of Network Science   21
        1.3.2 Networks and Network Research in the 21st Century . .   23
        1.3.3 Status and Challenges of Network Science   25

2 Basic Network Graph Models and Their Properties   29
   2.1 Graph Theory Fundamentals   30
        2.1.1 Basic Definitions and Notation   31
        2.1.2 Additional Definitions   35
        2.1.3 Connectivity   38
        2.1.4 Paths and Cycles   46
        2.1.5 Flow   48
        2.1.6 Planarity   50
        2.1.7 Coloring (Covering)   53
        2.1.8 Algebraic Graph Theory   55
   2.2 Random Graphs   59
        2.2.1 Basic Random Graph Models   60
   2.3 Notation   63

3 Cognitive Methods and Evolutionary Computing   65
   3.1 Brief History of Evolutionary Computing   66
   3.2 Elements from Evolution Theory   66
   3.3 Evolutionary Computing   68
        3.3.1 Components of Evolutionary Algorithms   70
        3.3.2 Representation   72
        3.3.3 Fitness Function   73

        3.3.4   Population   73
        3.3.5   Parent Selection   74
        3.3.6   Variation Operators: Recombination and Mutation . . .   74
        3.3.7   Survivor Selection   75
        3.3.8   Initialization and Termination Conditions   75
        3.3.9   Operation of Evolutionary Algorithm   76
   3.4 Evolutionary Computing Approaches   79
        3.4.1   Genetic Algorithms   79
        3.4.2   Evolutionary Strategy   91
        3.4.3   Genetic Programming   95
        3.4.4   Evolutionary Programming   98
        3.4.5   Evolutionary Computing at a Glance 100
        3.4.6   Parameter Control in Evolutionary Algorithms 100
        3.4.7   Special Forms of Evolution 105

4 **Complex and Social Network Analysis Metrics and**
   Features   **109**
   4.1 Degree Distribution 110
   4.2 Strength 113
   4.3 Average Path Length 114
   4.4 Clustering Coefficient 115
        4.4.1   Definition 115
        4.4.2   Extension to Weighted Graphs 116
        4.4.3   Extension to Directed Graphs 118
   4.5 Centrality 122
        4.5.1   Degree Centrality 123
        4.5.2   Closeness (Path) Centrality 125
        4.5.3   Betweenness Centrality 128
        4.5.4   Betweenness Centrality Approximation Methods 130
        4.5.5   Eigenvector Centrality 135
        4.5.6   Example of Centralities' Computation 136
   4.6 Prestige 138
        4.6.1   Degree Prestige 139
        4.6.2   Influence Domain 139
        4.6.3   Proximity Prestige 140
   4.7 Curvature 141
   4.8 Metrics at a Glance 143

5 **Distinctive Structure and Features of Complex Networks**   **145**
   5.1 Network Structure and Evolution 145
   5.2 Small-world Paradigm 146
        5.2.1   Prolegomena—Description of a Small-World Network . 146
        5.2.2   Large-scale Experiments—"Six Degrees of Separation" . 148
        5.2.3   Watts and Strogatz Model (WS Model) 150
        5.2.4   Kleinberg's Model 153

        5.2.5   Examples and Applications 155
  5.3   Scale-free Networks 158
        5.3.1   Definition and Properties 158
        5.3.2   Examples and Applications 161
        5.3.3   Barabási–Albert Model 162
        5.3.4   Extensions of the Barabási–Albert Model 168
  5.4   Hyperbolic Structure of Complex Networks 176
        5.4.1   Background on Hyperbolic Geometry 176
        5.4.2   Evolutionary Models Developed on Hyperbolic
              Geometry 178
  5.5   Expansion Properties 180
        5.5.1   Definition and Analytical Properties 180
        5.5.2   Applications of Expander Graphs 184
  5.6   Conclusions 185

6 Evolutionary Approaches   187
  6.1   A Brief Description of Wireless Multi-hop Communications . . 188
  6.2   Topology Control (TC) and Inverse Topology Control (iTC) . . 191
  6.3   Spatial Graphs and Small-World Phenomenon 192
  6.4   Inverse Topology Control-Based Approaches 195
        6.4.1   Early Approaches Using Wired Shortcuts 196
        6.4.2   Approaches Using Wireless Shortcuts 202
  6.5   Holistic Topology Modification Framework 208
        6.5.1   Weighted Edge Churn Framework 209
        6.5.2   Weighted Node Churn Framework 216
        6.5.3   Combined Mechanism (WEC and WNC) 220
        6.5.4   Optimization Methodology 221
  6.6   Special Cases 229
        6.6.1   Example 1: Elimination to Binary Graphs (SETM) . . . 229
        6.6.2   Example 2: Trust Management in Wireless Multi-Hop
              Networks 236
  6.7   Conclusions 244

7 Conclusion   247
  7.1   Lessons Learned 247
        7.1.1   Emerging Trends and Their Benefits 248
        7.1.2   Discussion on Evolutionary Topology Modification
              Mechanisms 251
  7.2   The Road Ahead 251
        7.2.1   Route Covered Already 251
        7.2.2   Open Problems 253
  7.3   Epilogue 255

Appendices   257

A Geometric Probability   **259**
  A.1 Probability Theory Elements 259
  A.2 Probabilistic Modeling of the Deployment of a Wireless Multi-
         Hop Network 261

B Semirings and Path Problems   **263**
  B.1 Monoids 263
  B.2 Semirings 264
  B.3 Examples 265

References   **267**

Author Index   **277**

Subject Index   **283**

Preface

In this book, we start the exploration of evolutionary dynamics for complex networks with a working example that might look familiar to many people, at least partially. The specific example intends to highlight the various aspects of everyday life and work where humans encounter, use, act on and obtain information from their interaction with complex communications networks, as well as numerous other emerging functions. Such an example will drive the following steps of the exploration of this book into a research domain currently shaping and even expanding rapidly. Then we provide a more elaborated overview of the topic, related goals, and features provided by this book. Finally, we provide potential alternative strategies for studying the material provided, depending on whether the book is intended as a research reference, an undergraduate or graduate course reference manual, or simply a reference of broader interest.

Working Example

An executive of a multi-national company hires a taxi from Tokyo Financial Center to go towards the International airport, to catch her return flight to the United States. The driver quickly consults the GPS device of the vehicle for the traffic street map, in order to identify the fastest route to the airport. The executive has requested a speedy lift, which will ensure she boards on time. The route duration estimation is 40min, safely enabling the executive to check her email through her mobile handheld device in the meantime, rather than worry about catching her flight. Both the accuracy of the GPS application and the luxury of a good international roaming plan allow her to focus on her job, while not wasting energy on logistics. The latest email requires her to complete a report, thus, she also opens up her tablet device, connects to the 4G network of the local provider, and via roaming and VPN, she connects safely to the intranet of her corporation, in order to complete the short report online and make it accessible to her supervisor in Philadelphia as soon as possible.

Incidentally, as soon as the taxi reaches the airport, the executive receives an emergency call, informing her that she needs to reschedule her flight and visit the corporate offices in Los Angeles for a couple of days before returning to the headquarters in Philadelphia. She heads straight to the airline helpdesk and requests an immediate flight change. The teller checks with the information system of the airline and is able to locate a convenient flight reroute. The executive will now fly from Tokyo to Chicago and instead of boarding the connecting flight towards Philadelphia, she will board a connection towards Los Angeles with only a slight delay of an additional hour. It just occurred to the executive how important the airliner network proves to be in such cases, where scheduling and other data have to be recalled and updated upon such last minute modifications. Next, the executive checks the new flight data, boarding times, duration, distance, mileage account, etc. from her smartphone via the airline application account. She also updates her social network profile status property in all social networks she uses, while waiting to board, in order to let her close friends and colleagues know of her coming plans.

During the flight the executive connects to the airplane's WiFi infrastructure in order to check her social accounts, read the news, and watch something On Demand, such as a movie, or missed episodes of her favorite TV series. Two of her college friends now living in Los Angeles have sent her invitations for dinner, once they were notified by her social network status that she will be in town. The executive checks her schedule and decides to join them as soon as her assignment there is over. At the same time, one of her friend's article posts in the social network has gained numerous "likes" and popularity, and she decides to check the article that coincidentally appeared in her favorite newspaper, before she continues watching the latest episode of her favorite TV series she unavoidably missed due to her original assignment.

Upon her arrival at Los Angeles she feels somewhat exhausted and probably a bit sick. Suspicious from various spreading news stories she heard over the last couple of days in some of the cities she visited through her trip, she quickly consults some of the latest medical blogs for her symptoms, while waiting for her pick-up. She eventually believes she might have got the latest flu from her business trip and heads directly to the nearest available hospital, which she found through her smartphone device and a relevant application. By collecting information from medical blogs and hospital databases, and based on symptom input by the user in conjunction to the selected town, the app is capable of suggesting such a hospital. In the hospital, the physicians that examine her consult the hospital records for recent virus alerts, and eventually they decide to proceed with some further lab tests and physical monitoring.

Eventually, the doctors inform her that she will need to receive medication and the prescribed antibiotics will take some time to spread through her immune system, a cell network going through her body, before she will start feeling better and be safely released. In the meantime, the corresponding agency of the Centers for Disease Control and Prevention in the hospital needs to know her exact travel plan, in order to obtain more accurate estimates of the flu's spreading dynamics. The epidemics control and prevention agency, which is informed from the hospital authorities, is interested in assessing the danger levels for a virus spread in the general population, and thus experts of the agency consult with the executive, inquiring about her traveling. Combining such data with other information obtained by talking to similar patients, they will become capable of reconstructing a propagation network for the specific virus and variations throughout the nation, obtaining the respective infection and recovery rates and assessing the importance of the situation. This will enable them to better track the flu propagation network and thus increase the efficacy of their countermeasures when needed.

Meanwhile, the executive is released after a few hours of close observation and an imminent improvement of her physical condition. The latest antibiotics employed by the doctors were created based on feedback similar to that obtained by epidemiologists consulting with the executive in previous cases. The antibiotics reacted in a targeted and rapid fashion in her blood and cell networks, allowing her to recover fast and luckily continue her planned occupation with the minimum possible delay and overhead. She is now capable of continuing her job, without suffering nasty symptoms and feeling assured she will not be endangered in the coming days.

After her assignment, she finally meets with friends from Los Angeles, as planned during her flight, most of which are active in the financial field as well. They decide to start with dinner and at the same time spend some time discussing the latest trends in their jobs and lives, while occasionally checking several facts, photos, and articles online through their mobile devices. By the end of their dinner, they decide to continue their night by having some more fun. They can solve this quickly by using the latest social applications. In a city like Los Angeles and with some aid by their smartphones they decide to quickly book online seats for a theater play they found interesting among those offered in the application, since some available seats were luckily still available. They completed the transaction smoothly and rapidly, securing their spots in the theatre. They also came across a post in an online nightlife guide for a seemingly nice bar in the neighborhood, and decide to visit it right after the play. While in the cab, the executive reflects once more on how easy it has become to have all those options and features through the established networks and how their interconnection enabled them to do so much on such a short notice, whereas in former times, the same arrangements would take at least 1–2 days of prior arrangements.

A couple of days later, after all her activities in Los Angeles are finished and while on her return flight to Philadelphia, the executive is now more relaxed and feels like scheduling her weekend online using her tablet and exploiting the features of business class seating, like Internet access–communication capabilities. It is time to allocate some time for herself. She has received an invitation for a tennis match and she is able to confirm her availability through email and also book a court through the Web service of the tennis club where she is registered to avoid suffering any court unavailability. She was also informed of an upcoming birthday party of one of her close friends. She immediately indicated her attendance and in addition, she was able to chat with some other friends that were online for the proper organization of the party through her social network mobile service.

Having set her social activities for the weekend, while in flight she decided to selectively read the latest local news in Philadelphia, where it turned out several things happened while she was out of town. Being registered in customized and automated newsfeeds, she was able to quickly locate the most interesting articles for her taste. She was also able to quickly form her personal opinion on them and post some comments in the relevant discussion forums and blogs. She was even able to follow several immediate responses to her comments by spending the remaining flight time productively on topics she enjoyed discussing.

During a reflective moment, she realized she will not be required to manually update all of her updates and modifications in her schedule and document management systems, as the latest cloud services will do so automatically, allowing her to find everything in order once she turns her devices on back in her place.

As soon as she landed, she was relieved that yet another demanding business trip came to an end. While riding a taxi from the airport towards her apartment, she spent some time reflecting on various moments from her trip and it occurred to her how many times she used modern technology and especially networked devices and services in order to finish her job more easily, communicate with colleagues and friends, and eventually document and boost her work. Once more, she had to appreciate the benefit of a networked lifetime. That eventually made her think a bit more how many other aspects of networked life she was actually using on a daily or less frequent basis and in a subconscious manner. All these have been delicately underlaying her daily routine for quite some time, and those regarding her health, from the moment she was born. She felt a bit happier her everyday life was now easier than what it would have been decades back and felt calm that she was now in a position to get some rest, enjoying what seemed to be a very revitalizing and fascinating weekend before a new week in the job started.

Topics and Features of the Book

The example of the previous section cannot characterize by any means a typical person, even in the more technologically-aware societies of our contemporary world. Even though there exist people that have to cope with similar rigor, most of the professionals and non-professionals have a more plain style of living. However, the example presents cumulative various facets of modern societies with respect to working, socializing, hobbies, entertainment and practically any aspect of human life. It can be suggested that the example combines many of the challenges that numerous people have to face, various other tasks they have to accomplish, or occupations they want to achieve on a daily basis. It illustrates how modern technology has changed human life, and how many different tasks can be possibly achieved nowadays with the use of modern devices, services, and infrastructures, even under strict or emergency constraints. Furthermore, the example provides a glimpse into the future demands and requirements that the modern style of living might impose over communications technology and information societies in general.

Above all, the example highlights the important role that various types of networks play in our modern societies. Essentially, it illustrates how modern societies are centered around various types of information and infrastructure networks. These networks span not only technological communications networks, such as those formed by mobile devices, the Internet, airline databases, and mobile applications, but also involve other type of networks as well. For instance, biological networks, such as protein and nerve networks, virus infections, financial networks, such as stock and trade markets, and numerous others. Networks of different types emerge everywhere, and conversely, similar types of networks emerge in rather diverse fields of human activity. For example, small-world networks, which will be analyzed in detail in subsequent chapters of the book, emerge in biology, computer networks, and social networks, at the same or different scales.

Most people would identify a subset of the presented example to match their daily lives to a lesser or greater degree. Through that, one may identify a plethora of naturally emerging networks or developed networks that have been specifically designed to interfere with our daily lives to the extent that this becomes routine. In any case, the emerging networking structures are very important and in addition, they operate most of the time in the background, as it happens with biological, financial, and other critical types of networks. What is more important is that this trend of identifying and exploiting more consciously various types of networks is increasing with a strong tendency to further intensify. As the technological means enable researchers and professionals to perform larger scale studies with greater accuracy, better control of such network structures and mechanisms developing on them can be achieved and eventually exploited for improving the quality of living of the current and future generations.

Among others, the above example illustrates characteristic cases of human interaction and emergence of various and diverse types of networks in our modern lives. Starting with telecommunications, professionals are able to perform their jobs remotely from various places of the world, as if locally present in their offices. Travelers are capable of scheduling and adapting their journeys on the fly and in the most efficient manner by exploiting mobile devices, global services, and integrated infrastructure networks. Airlines are in a position to reschedule their passengers, ensuring the most efficient transportation of persons and products in the most convenient manner. Doctors and epidemiologists are capable of monitoring the spreading of diseases and viruses and eventually may determine the severity of a threatening situation for the population. Also, the understanding of biological and metabolic pathways within the human or animal cell networks enables the development of more efficient and more rapidly acting medicines. Critical or more casual information on financial markets can also be quickly communicated through hybrid social-communication networks, enabling more educated decisions and safer risk asset management, or learning about rumors and critical updates that could optimize decision-making and minimize the undertaken risk. Finally, modern communication and IT infrastructures have enabled the development of various social networks, which in turn have enabled people to experience various events, updates, and news even when far away from their friends, families, or other social groups they belong to. Such social networks have enabled the integration of social circles with activities in a virtual reality that could be properly exploited for a more exciting and valuable daily life of modern societies.

The common denominator and main feature of all the above emerging cases of networks in all aspects considered is the formation/development, operation, control, and eventual exploitation of the emerging networking structures, either natural, human-initiated, or completely artificially engineered. All such networks interfere directly with our lives in various capacities and for various purposes. Several of them appear simple in their operation, while others yield such complicated behaviors that currently we are not even close to understanding them, setting aside any aspirations for obtaining some type of desired control over them. This book will consider exactly this feature of the emerging networks and more specifically it will do so from the perspective of evolutionary dynamics, namely by focusing on the factors and mechanisms affecting the dynamic modification, spontaneous or designed, of these structures, as they evolve in time and sometimes even in space.

The second notable feature in the example of the previous section is the evident diversity of the emerging networks and corresponding processes/mechanisms developing on top of them. The application framework diversity where these networking structures develop is even more diverse in scope, nature, and objectives. However, in most cases even within this diversity, fundamental, common, and generic problems of a mathematical nature emerge among the various disciplines. For instance, the networking problems of virus propagation among host machines in a wireless decentralized communications network is similar in nature and mechanics to the problem of virus propagation in humans, faults in engineering complex production lines, and news in information channels around the globe. This potential provides a great opportunity for tackling critical problems in generic, efficient, and convenient ways that would benefit multiple disciplines cumulatively and magnify the potential progress in various fields. This book will cover the diversity exhibited by the modern study of networks, by presenting and analyzing the available networking structures along with their properties, applications, and special features. Even though the book will not be exhaustive, the most characteristic, important, and useful network types with applications in the most interesting fields are included in the analysis provided.

The main topic of this book is devoted to the objective of first analyzing complex communications networks by exploiting multiple and diverse mathematical methodologies from other disciplines of Network Science and then improving their operation and control approaches over them in a seamless and efficient manner. This is a twofold goal, which will start by introducing the basic analytical tools that can be used for the analysis and study of complex networks in general, independently from specific scientific fields of research. Following this, the focus will shift to exploiting fundamental elements of complex networks in more advanced evolutionary mechanisms, which can be employed for improving artificial networks and more specifically wireless decentralized communications networks, such as ad hoc, sensor, and mesh networks. The whole approach taken is a holistic study of complex communications networks inspired by the inherently hierarchical structuring of more general complex networks themselves (as will be explained in detail in the next chapter) and the goal will be to jointly exploit elements from the multiple perspectives and mechanisms of complex networks in the improvement and advancement of communications networks, in a manner that will enable and inspire similar efforts in other disciplines as well and thus, eventually, contribute to developing a novel and constant-improvement feedback approach among the various disciplines of Network Science and complex networks, as will be presented in the following chapters.

Roadmap and Book Objectives

This book has mainly been developed as a self-contained volume covering both breadth and depth of evolutionary dynamics for wireless communications networks. However, multiple uses may be achieved by following different coverage sequences and selecting different parts of each chapter.

Following the instigated chapter sequence provides a gradual and holistic viewpoint of the corresponding field, starting with coverage of the broader scope of Network Science and complex networks, then proceeding with the development of the necessary background in terms of mathematical content and algorithmic approaches, and finally, progressively applying several of the concepts presented in the background-devoted section into frameworks and mechanisms covering the dynamic and evolutionary behavior/improvement of wireless networks.

In the aforementioned thread of coverage, several parts have been covered in more detail, while others have been briefly touched on, citing other sources that the more interested audience may consider. In case the reader wishes to obtain a holistic perspective of evolutionary dynamics, but at the same time does not have the luxury of time, several more specialized parts have been noted with an asterisk and could be omitted in an initial, more breadth-oriented study.

The book also jointly covers the theory and applications of the presented approaches. The reader may select those theoretical approaches or applications that are of more interest or relevance to his/her interests on an on-

demand basis and according to the level of detail he/she wishes. References
to more extensive sources and manuals or more detailed treatises are provided,
when the treatment in the book does not contain significant detail.

This book is also meant to be used in other ways, apart from its basic reference manual use. One of these possible uses is as an undergraduate textbook
for more advanced relevant courses, e.g., introductory courses on Network
Science, complex networks, or evolutionary network dynamics. Some basic
knowledge of computer/communications networks probability theory and differential equations is considered prerequisite in these cases.

Finally, another use, and perhaps a more targeted one, is as a main textbook for an introductory graduate course, covering the fundamentals of complex network analysis and network engineering. In this case, the book can
be used to prove theoretical tools and practical examples for eventually exploiting elements of complex network analysis in the design and optimization
of communication networks, and thus aid students working in relevant fields.
However, the methodologies presented could be of interest for graduate students and researchers of other disciplines, i.e., social sciences, etc., in which
case the book can be exploited in a selective manner, where the instructor
will be choosing excerpts of interest in a more focused manner.

As such, and since the complete material cannot be covered within a single
semester course, both at the undergraduate and graduate levels, we suggest
potential chapter layouts that could be utilized for constructing coherent material for relevant courses. The flowchart in Figure 1 depicts such chapter out-
lines. In the middle, the cover-to-cover chapter flow approach is shown, as the
main suggested coverage of the book. On the left and right hand sides of the
figure, the flowchart depicts suggestions for deviating from the cover-to-cover
flow of the coverage, which can be determined according to the specific objectives of an instructor, for an undergraduate and graduate course respectively
(left-hand side for undergraduate and right-hand side for graduate uses).

Finally, we note that the book contains some appendices, which can be
used within a course as needed, or simply as references for the independent
researcher or student, should a quick reference background be needed, as noted
in the text.

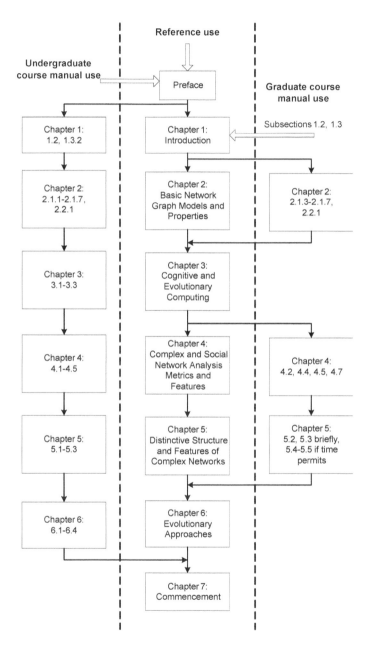

Figure 1: Potential use of the book as a long-term research companion and/or undergraduate or graduate course manual.

Chapter 1

Introduction

A constant drop in computing costs can be observed in consumer markets [113], realized both in terms of cheaper devices with more capabilities at lower prices (e.g. cheaper cellphones, smartphones, tablets more capable than their desktop counterparts five years ago, etc. [85]) and in terms of massive and publicly accessible computing infrastructures (e.g. cloud services, commodity virtual machines, public services, etc. [141]), and this has made pervasive communication and information exchange massively available as well [138]. It has essentially turned computing power into a commodity, available to the industry and individual users alike and at various available scales (from small businesses to large production lines and home applications) [58]. In turn, this trend has shifted the focus of technological evolution to the effective interconnection of widespread communicating computing units, rather than their further evolution, drawing even more attention to the networking aspects of the research and technology of such devices/systems. Of course, this does not mean that research and technological progress in computing resources has diminished. However, the greatest interest of the industrial and research communities has now shifted more towards the exploitation of computing resources, which most frequently involves the Internet and relevant infrastructure/access networks for exploiting the available commodity computational infrastructures. Consequently, it may come as no surprise that the inter-networking issue of computing has emerged as one of the most prominent pillars of the intersection between two relevant fields of science, namely computer science and electrical engineering, the first one covering the computing aspects and the latter the networking aspects of the aforementioned systems.

This book focuses on the more advanced aspects of the emerging networking problems, relevant to the trend described above, and spans a number of various and diverse topics centered on the more general field of complex networks and the branch of Network Science that covers the emergence of networks from a broader perspective. The terms complex networks and Network

Science have been coined to describe inherently inter-disciplinary research establishments that concentrate around the concept of network, irrespective of the applications and situations where this notion emerges. Network Science and complex networks constitute a vast field of science, in which multiple and diverse research facets emerge, many of which will be described in a detailed manner in the following subsections and chapters. It should be noted though that in principle, this book leans more towards the telecommunications aspects of networking, as it mainly provides examples of such networks. However, it explores inter-connections and relations of methodologies/mathematical models that have been developed in disciplines other than telecommunications, or can be extrapolated from telecommunications into other disciplines as well.

In order to achieve the above goal, the book follows a hierarchical approach, in which the required knowledge will be gradually developed, enabling the understanding and analysis of more advanced concepts and methodologies. However, for completeness purposes and in order to assist the non-expert reader, the book is also involved in the presentation of the required background before delving into the study and development of evolutionary methodologies and analytical tools. This chapter serves as an introduction to the broader areas spanned by the book. It serves the purpose of explaining the notion of complex networks and in the broader sense that of Network Science.

The presentation of complex networks in Section 1.2 is a stepping stone, which will develop all the fundamental notions relevant to networks and their operation in general. Also, it provides the bigger picture for the applications, yielding potential directions for further studies and uses of the proposed-problems/solutions. Secondly, Section 1.3 attempts to provide further indications and insights both for the presented topics, as well as for other relevant or implicitly relevant topics that emerge from the analysis of those included in this book. Lastly, this chapter provides a concise summary of the objectives that the whole book aspires to cover. It highlights several of the features presented, as well as the framework within which such features will be considered, e.g., the wireless communications paradigm within which some of the standard analytical tools employed in other disciplines have to be considerably modified, using radical approaches, in order to overcome provenly tough problems.

1.1   Approach and Objectives

As already mentioned, the main focus of this book is on the study and analysis of methodologies and models that have been developed within the framework of complex networks and Network Science for communications networks. By exploiting multi-/inter-disciplinary features emerging within this framework, the book aims at the improvement of the design, operation, and performance of the latter types of networks at large. In order to better demonstrate the broader concepts introduced and presented regarding the exploitation of Network Science elements in computer/communication network design and analysis, the book will adopt the perspective of wireless networks. Thus, the main objective related to the above scope will be to present and then exploit analytical methodologies from different disciplines of Network Science and complex networks in the design and more importantly in the control of wireless networks.

Traditionally, wireless communications networks have been operating in stringent environments and frequently under varying conditions (other than environmental, e.g., user or mission related). Methodologies that were well-established, or at least efficiently working, for their wired counterparts, have been proven oftentimes at least inappropriate and occasionally, not even viable. Routing is a characteristic representative aspect of this fact. Traditional routing designed for the Internet and local area wired networks did not scale or operate well in wireless ones [43], leading to the design of new wireless-specific routing protocols [140], which, however, were not completely successful for their own part. For instance, several performance and implementation issues remain, most of which owe their existence to their wired counterparts, e.g., scaling of the routing approach [42].

For these reasons, the latest trends in user demand, applications, and services, combined with increasing volume of traffic and technology paradigm shifts, have created high expectations for more radical approaches in the design and control of wireless infrastructures. This book focuses precisely on introducing novel and radical methodologies for designing and dynamically controlling the wireless networks of the future. In this effort, the benefit in the engineering approach to observe, understand, and exploit features from disciplines related to networking will become apparent, thus developing novel radical analysis and design tools. It will also become apparent how straightforward it will be to follow the other direction as well, and thus enable benefits for other disciplines as well through this two-way approach.

The book follows a hierarchical approach, based on which it first provides the required background for obtaining a solid mathematical language/notation, thus enabling the easier identification of emerging models from the observed network behaviors. Such background in this case includes mainly traditional graph theory and the latest random graph theory basics. The first is focused towards deterministic and static networking paradigms, while the latter is oriented towards dynamic and varying behaviors. Then, more advanced models that focus on the evolutionary behavior/optimization of networks will be studied.

Apart from the fundamental background knowledge, the approach of the book will be involved with methodologies that enable radically improving the traditional architectures and operations. Introducing elements from complex networks and Network Science in the analysis of communications networks, mainly exploiting them in the fields of variational processes, control, and distributed computing, is a key objective of this book, one that not only provides new perspectives in the analysis of communications networks, but also enables the reconsideration and evolution of such networks.

Another important goal of this book will be to provide paradigms and methodologies that can be more holistically applied in other disciplines of Network Science as well, possibly after small adaptations depending on the specific discipline that the extrapolation is intended for. As will be more analytically explained in Section 1.3, this is a more far reaching objective, given the current status of achievements in Network Science. However, this book will attempt to provide a small-scale approach that could be extended to a broader methodology (at least regarding its strategic steps) for transferring analysis approaches from communications to other types of networks.

This book devotes special attention to evolutionary network design and control, which mainly covers the dynamic behavior exhibited by communications and other types of networks. One of the most fascinating features of network evolutionary dynamics is the fact that in many cases, networks change from being totally random to being completely symmetrical and predictable, a fact that has been proven to have great importance for human sustainment (if one considers the corresponding biological networks exhibiting such behavior). Order emerges out of chaos and sometimes the converse takes place as well, through completely individual actions or collaborative behaviors. Several such examples include various application frameworks, from the electrons in a superconductor to the pacemaker cells in our hearts. And even though these phenomena might seem unrelated on the surface, at a deeper level there is a connection, forged by the unifying power of mathematics, and especially regarding networks by the mathematics underlying Network Science [152].

Summarizing the main and secondary objectives of this book and the employed hierarchical approach, the following list contains the goal-topics to be covered in the remaining chapters of this book:

1. Study the methodologies and features of complex networks and Network Science that could be of potential use in communications networks.

2. Study the evolutionary behavior of communications networks by observing trends already analyzed in other disciplines of Network Science.

3. Introduce novel design and control methods for communications networks, inspired by social networks and other complex networks.

4. Create holistic frameworks and methodologies that can be applied in various disciplines of Network Science with minimal adaptations.

The approach adopted will cover the above subjects in a generic manner, potentially accommodating multiple perspectives for each approach. Namely, the presented frameworks and methodologies can be applied in various types of networks and their applications. However, the main perspective followed will be that of wireless communications networks, and most of the examples, problem instances, and solutions presented will be drawn from the area of wireless complex networks.

1.2 Fundamentals of Complex Networks

The interest of the research and industrial communities, as well as the public sense for networks, has grown substantially, especially in the last decade. Their drastic and vast proliferation and technological penetration has increased the observed public awareness in any type of form in which networks emerge. Networks are nowadays omnipresent and have been identified as highly crucial in most of their application frameworks. For example, protein interaction networks in biology, which essentially implement a molecule signaling and control toolbox, are highly important for human body operation. In computer networks and telecommunications, mobile networks have enabled pervasive communication between people across continents, diverse conditions, and degrees of importance and criticality.

Similar observations to the above, and many more others observed in natural and daily social lives, lead to the fact that if one needs to characterize modern societies in a couple of words, these terms must be *connected*, *interconnected*, and *inter-dependent*. In addition, even though these terms represent the bigger picture only from a narrow perspective, they are very successful in providing the essential elements dominating our everyday lives.

Another key observation of emerging networks in our lives, widely accepted in both the research and industrial communities, is that the complexity of most interconnected (inter-networked) systems is not in the behavior/operation of a single unit or larger component among the many constituting a realistic system, but rather in the cumulative behavior/operation exhibited by the interconnection and communication of such individual units. Namely, the inter-networking of such units/modules is more important for achieving more beneficial analysis and control of such systems at a lower cost. Such inter-dependence of nodes is closely related to the notion of collaboration as well, where nodes might be working together for achieving a more complex objective that would be otherwise impossible to achieve individually by a single network entity. Inter-dependence and collaboration are critical aspects of almost all types of networks, and will be widely considered in the rest of the chapters of this book.

The main research efforts in the past were centered in the understanding and analysis of the behavior of individuals units and components of them and the achieved progress has been fascinating. However, as the level of understanding increases and daily demands for added-value knowledge and services increase, the inter-dependent behavior of such basic modules gains interest and sometimes it becomes essential in order to achieve the desired level of control and flexibility over these structures.

Researchers involved in different capacities in the study of emerging networks have lately used a new term, namely complex networks, in order to cumulatively refer to all network research in diverse and multiple disciplines, namely [6], [115].

Definition 1 *(Complex networks) A complex network is one that exhibits* *emergent behaviors that cannot be predicted a priori from known properties of the network's constituents.*

The above definition does not explain the notion of a network, which will be analyzed in more detail in the following subsection. It rather focuses on the characterization of networks as "complex." It considers especially that the observed behaviors can be diverse and completely different, even within the same discipline (e.g., social networks), but could also exhibit unexpected similarities even when observed across diverse disciplines (e.g., malware propagation in wireless multihop networks resembling the virus propagation in animal species or humans).

The corresponding field of complex networks covers a very broad span of network types and emerging features, problems, and mechanisms of broader scientific interest. The proper definition, study, and classification of the corresponding diverse types of involved network structures will provide a solid basis for identifying and revealing common emerging problems. It will also reveal the most suitable approaches that can be exploited from different scientific fields, i.e., other than complex communications networks (such as systems biology, finance, and sociology), in order to address the involved problems more efficiently and develop more feasible/practical solutions.

By considering the various networks involved in the study of complex networks and their emerging behaviors, in general, three dominating features are characteristically observed. The first one is that when it comes to modeling the interactions between network elements, this is achieved by links connecting these entities. Links may represent various forms of interactions or relations. The second element is that nodes exchange different types of resources across such links. The resources can be rather diverse, e.g., in communications networks, nodes exchange data in bit form, protein networks exchange aminoacids, pipeline networks transfer various types of fluids (blood for veins, oil for oil pipes, water for water utility networks), and others. As before, the resources could be of different and diverse natures and quantities and convey diverse meanings. Finally, the third element is that nodes interact through the direct links defined. Thus, two nodes cannot interact directly unless they share a common link between them. This type of interaction refers to a physical inter-connection of the nodes linked directly, e.g., two people being best friends in a group of humans. However, interactions can also be implicit, for instance, in the case of two people being linked to each other in a social network, without physically knowing each other, but rather simply because they shared another third link with a person who happened to be friend with both in reality. Thus, in this book we are mainly involved with direct interactions between network entities, especially in the more specific parts of the presented frameworks. In any other case that interactions are implicit, this is explicitly noted and relevant considerations are made.

Regarding the current information-based and network-dependent societies, another two prominent features may be observed. The first is the diversity of the emerging network structures arising almost holistically in every application aspect that one could possibly think of. For instance, network structures emerge in biology, societies, engineering, nature, and practically all other aspects of artificial operation (initiated and dominated by humans). The second is that such networks consist of participating entities of various potentials and intelligences. Different computational and decision-making capabilities by such entities lead to nonpredictable cumulative behaviors. The latter is the main reason that the corresponding interest has risen so much lately and has attracted the attention of various scientific disciplines.

It has been observed and in some cases formally quantified, e.g., in communications networks, that most of the important networks surrounding us are becoming larger, increasing in the scales of millions or billions of users/actors/players/entities/etc. Some of these characteristic networks in accordance with their projected current order[1] estimate (as documented in various sources of the bibliography and the Internet) are provided in Table 1.1.

A more complete network taxonomy is provided in subsection 1.2.2, along with other features they exhibit, revealing more facets of complex network diversity.

In this book, we emphasize networking aspects that are of a decentralized nature and their structure and behavior resembles that of most distributed complex networks encountered in general. Furthermore, the proliferation of computing devices and computers in addition to the development of diverse and social networking applications has led to the emergence of a new and rapidly developing application area, namely that of online social networks. The social dimension has been shown to have significant impact on wired networks, especially the Internet [135], and it is expected to have a similar if not a more vast one on wireless networks, due to the capabilities of modern smartphones and respective provider services. Thus, in order to provide a more complete picture of the methods associated with analyzing and controlling complex communications networks, we also focus on the impact of the social network layer on the actual physical one.

This book takes a more radical perspective, by establishing a top-bottom approach in addition to the traditional bottom-up. In the latter, in complex communications networks, most of the network design techniques took an approach where the lower protocol layer mechanisms affected the design of the higher ones. For instance, the properties and operation of the physical layer have been taken into account for the design of the MAC layer protocol and the MAC protocol has been considered for the design of the routing functions in turn. However, this approach has not yielded fascinating results in the case of wireless distributed networks. In most cases, the employed mechanisms are essentially the ones designed for wired networks, properly adapted in order to

---

[1] The order of a network denoted in the caption of the figure as well is formally defined in Chapter 2, Section 2.1.1.

Table 1.1: Scaling of order of various complex networks.

Network	Order of scale
computer networks	billions
Internet	billions
corporate network	thousands
home network	dozens
university campus network	thousands
cellular phone networks	billions
electrical power grids	trillions
sensor networks	hundreds of thousands
roadmap networks	trillions
social networks	billions
food webs	hundreds
brain cell networks	billions
protein interaction networks	hundreds of thousands
affiliation networks	hundreds
citation networks	hundreds
open market networks	decades
bank networks	thousands
GDP^2 flow networks	hundreds
cash flow networks	billions
air-traffic networks	thousands
collaborator networks	decades (hundreds)
logistics networks	thousands–millions

yield the desired operation in wireless. Quality of Service (QoS) and advanced features, such as good scaling, resilience, etc., were not considered due to the fact that requirements for such elements were not realistic in the early days of emergence of distributed wireless networks.

However, as the requirements and demand for such advanced services increase every day and modern applications have become the driving force, rather than the underlying technology, more sophisticated topology modification mechanisms are required. According to the approach of this book, elements of the higher layer, such as the social, are exploited for directly modifying the lowest physical topology, thus closing the design loop in an evolutionary fashion, similar to the one observed in natural cognitive processes [109]. Such an approach will allow more targeted and efficient adaptations of the underlying complex communication network topology, thus increasing the value of an infrastructure without requiring major cost/resource sacrifice.

The following chapters (Chapter 2 and Chapter 3) will focus on the background theory, as well as the classification and analysis of various network

---

^{2}GDP: Gross domestic product.

modification mechanisms for wireless decentralized networks that exploit social features from the corresponding online social networks. The engineering of complex networks of any type is not predictable and/or controllable because the scientific basis for analyzing, building, and evaluating such designs is still immature. Thus, getting a grip of the fundamental science of networks in terms of structure, dynamics, and evolution is a topic of immense interest and critical value for the benefit and progress of human societies, as covered in this work.

The level of analysis in complex networks spans multiple and diverse perspectives for different types of networks. For instance, in complex communications networks, there are mainly three analysis perspectives, i.e., physical, logical, and social, as explained before and as will be analyzed in more detail in the following chapters. Fully understanding such analysis perspectives will enable building cross-level mechanisms in a cognitive fashion for each network application, in a manner where not only the mechanisms of a layer build on the features of lower-layer mechanisms, but also the lower-layer mechanisms exploit features offered by the higher-layer mechanisms. This book will offer the background and arsenal to achieve such cognitive operation in complex networks, with special emphasis on wireless complex communications networks.

Finally, we refer to a significant aspect of complex network analysis, which refers to the modeling of different network types represented or studied analytically, or their modeling as nodes bearing a specific processing (differential) rule, etc. In the first, a complex network represents the diversity of the various types of networks included and treated cumulatively as network models. In this case, the analysis takes into account that the property is studied across the various network types considered, which exhibit different properties and topologies. Thus, in order to consider such variations, for instance studying routing in ad hoc and cellular networks, a generic 'complex' network type is considered. The term complex on this occasion is indicative of the various and diverse properties exhibited by the different topologies needed to be taken into account in the generic study of a process, i.e., here, routing. In the second case, the term complex refers to the actual nodes of a network, which can also vary in application and scope, and characterizes the complexity of their features. More specifically, complex network nodes may vary in intelligence and processing capabilities. However, if nodes are capable of executing some type of computation, simple or more advanced, then the cumulative behavior could exhibit various degrees of complexity and, thus, complex behavior may be observed or engineered on demand.

The first step towards understanding such emerging and complex behaviors is to understand the fundamentals of network emergence and operations and then deal with the tools required for their proper control. In the following we start with network fundamentals and their network taxonomy, in order to provide a concise overview of networks and their application span.

1.2.1   Complex Networks Fundamentals

The cornerstone question of complex networks regards the overall reason for the formation and emergence of networked structures, within any type of application framework and diverse operation of the networking structures. This question has become more prominent lately with the increased interest attracted by complex networks. It has been more broadly and thoroughly put across the disciplines involved in the theory of complex networks than in the past.

Interest in network research has exploded during the past ten years (especially the last ten years for communications and the last five for a broader interest in different types of network research). Networks enable the necessities and conveniences of modern life, which can be easily observed in multiple facets of human social life and natural processes. For instance, different types of communications networks enable diverse and pervasive types of communications among people, online social networks have enabled new forms of social contact and new norms of social living, and transportation networks of different types, e.g., international highways, air-corridors, etc., have enabled more efficiency in terms of time and consumed resources in the transportation of people and goods. Especially, engineered networks are a major driver of the increasingly global economy and social evolution, as can be verified with the cases of road networks, air traffic airways, telecommunications networks, and, lately, online social networks.

In any case, scientists involved in various capacities in the study of emerging networks, and especially those with analytical backgrounds, have wondered whether there exists a single and broad reason explaining the formation of networks across all different application perspectives. The importance of such a reason would be significant since it would not only explain why networks develop in various facets of life, but it would also drive the evolution of such networks and indicate the dynamics of their typical behavior.

Substantial consideration has been accumulated on this key question and across disciplines. It has turned out that the answer to this critical question is a simple and profound, yet critical emerging trade-off underlying the existence of all networks and involving the operation of all the entities constituting a network. More formally:

Definition 2 *(Network formation) The main reason for the formation of any network observed in any aspect of nature or human society is the emerging trade-off of gain versus cost of collaboration for the entities constituting the network with their (inter)-relations or the network cumulatively.*

The whole concept of such a trade-off is based on the notion of collaborative operation. In fact, collaboration of network entities appears as the fundamental reason for the formation of a network and the gain (benefit) or cost respectively emerge as consequences, namely measurable outcomes that drive the very reason of existence of a network (i.e., collaboration) to one or

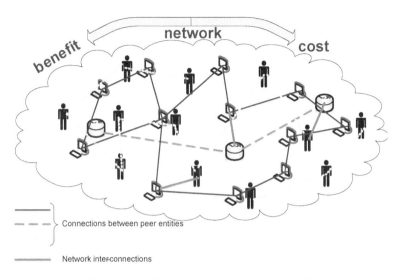

Figure 1.1: Network formation trade-off.

the other direction as shown in Figure 1.1. This notion of collaboration may seem weird, especially for distributed complex networks, where the entities are supposed to act selfishly, aiming only at their own benefit. However, even in this case some sort of collaboration between network entities is implicit (or underlying), so that the whole structure remains concrete. In a different case, should the network nodes be completely reluctant to collaborate in some sense, those nodes would sustain their operation without any need for a network structure. Their objectives would be better met away from the rest of the network, and if all nodes had such potential, there would be no point in forming a network. Thus, some form of collaboration is the very essence of network formation.

Definition 2 does not only mention collaboration as the reason for network formation. In addition, it describes the dynamics that drive network formation. Such mechanics are described by the underlying trade-off between benefit vs. cost of collaboration, which can be observed and quantified with measurable ways in all types of emerging networks and within all application perspectives where they appear. Furthermore, the benefit vs. cost of collaboration trade-off involves both the individual entities of the network, i.e., nodes, players, actors, etc., as well as the network as a whole and can be considered in various levels of the network, i.e., node level, node group level, or the whole network, by using different and diverse quantitative parameters. A network engineer can exploit the benefit vs. cost trade-off to drive the design of various mechanisms of the networks from one direction of the trade-off to the other depending on the desired operational requirements and application demands. For instance, in wireless ad hoc networks, the obtained benefit is the wireless and flexible transmission of data at the cost of energy consumed for the wireless transmission.

Table 1.2: Indicative examples of the benefit vs. cost of collaboration trade-off for various network types.

Network	Benefit of collaboration	Cost of collaboration
Computer networks	information (data) exchange	power/financial
Cellular networks	human communication/ mobility	power/financial
Internet	information exchange	financial
Wireless ad hoc networks	communication/flexibility	energy/interference
Cortex cell networks	message passing	energy
protein interaction networks	information tranfer/energy	food consumption
metabolic networks	energy	biochemical reactions
food webs	survival	effort
social networks	communications	privacy
air-traffic networks	resource management	communication cost
roadmap networks	shortest routes	initial and maintenance cost
affiliation network	context information	privacy
financial networks	information	privacy
GDP flow networks	economy assessment	independence
bank networks	cash flow monitoring	fraud risk

Table 1.2 provides indicative examples of the aforementioned benefit vs. cost of collaboration trade-off for various types of complex networks. The benefit and cost columns provide only some indicative benefit/cost types in each case, and in most cases, several other important benefit vs. cost trade-offs may be identified at various levels of analysis of such networks.

1.2.2   Complex Network Taxonomy and Examples

It has already been explained that complex networks are quite diverse, varying in numerous aspects of their structure, operation, and application framework/scope. For this reason, several classifications emerge, each of which is developed according to different metrics/features, while also serving a different purpose. In Table 1.3 we first provide a list of the parameters that can be used in different cases to segregate complex networks. Each parameter yields a different classification, so that a complex network might belong to different classes based on these parameters. For instance, a biological network may belong to the same class as a communication network according to the type of network, but the two might belong in different classes according

Table 1.3: Complex network taxonomy classification parameters.

Parameter	Means of verification
Network type	topology/implementation
Scope	operation/benefit
Application framework	environment/purpose
Origin (formation)	observation/simulation
Scientific discipline	mathematical analysis tools
Operation	developing mechanisms
Scale	number of entities/interactions
Performance	properties/features quantities
Reliability	network mechanisms
Security	topology/mechanisms

to the formation criterion. Network type is concerned with the very essence of each network in terms of entities-interactions, where different types exhibit different properties, features, etc. For example, a small-world network is fundamentally different from a random network.[3] The scope and application framework are slightly related; however, the first is concerned mainly with the benefit obtained in the gain vs. cost trade-off, while the second criterion is mainly concerned with the applications where the scope is realized, e.g., in communications networks the application framework spans over information dissemination, Internet services, etc. The origin criterion explicitly concerns the gain vs. cost trade-off and how this develops and leads to the formation of a network, while the scientific discipline cumulatively describes the mathematical tools that can be used to analyze and further control them. The operation criterion regards the developing mechanisms, mainly in algorithmic terms and their function/results, while the scale criterion is concerned with the variability of network behavior, as the size of the network in terms of entities and interactions between them varies. The latter is also closely related to the performance criterion, which can be used to classify networks, according to specifically defined performance indices. Finally, reliability and security are two criteria that have been employed lately for segregating the capabilities of different networks and this is taken into account further in the network design and analysis.

The second column of Table 1.3 denotes the means of verification for each classification parameter provided, namely how the classification can be realized. Specifically, regarding network types, the topology (in a mathematical sense) can be a differentiation factor, while for the scope criterion, the operation and benefit obtained by it (in the benefit vs. cost of collaboration trade-off) can be used for identifying and classifying the various networks.

---

[3]Both network types will be explicitly defined in the following chapters.

Table 1.4: Origin (formation) of complex network classification.

Natural	Human-initiated	Artificial
biological	social networks	computers
brain/cortex networks	power-law	mobile devices
evolutionary	online social networks	cellphones
genetic	business	sensor
transcriptional	open-market	delay-tolerant
immuno-suppressive	corporate	mesh
neuron networks	GDP flow	vehicular
ecologic	production	roadmap
protein networks	supply/logistics	air-traffic
regulatory	scientific	power-grid
substance networks	affiliation	artificial neural networks
material networks	family	IT networks
species	language	pipelines/utility networks
virus/disease	malware[4]	circuits
food webs	newsfeeds	transportation

Following the classification criteria, we then proceed with some interesting complex network taxonomies that serve various purposes and are of major interest for the scope of this book. Of course, the classification parameter list provided is non-exhaustive and many more criteria may be devised. Table 1.3 summarizes the most common ones with regard to current studies and concerns related to complex networks and Network Science study.

One of the most useful of these classifications is the one that takes into account the origin of network formation and operation, namely whether the network was formed spontaneously or artificially and whether its operation is dictated by natural or artificial factors as well. According to this parameter, complex networks maybe characterized as *natural, human-initiated*, and *artificial*.

Natural networks include those complex networks that emerge in Nature and continue their spontaneous operation/evolution for the duration of their lifetime. Characteristic examples are complex networks emerging in biology, such as protein receptor networks, blood cell networks, cortex neural networks, etc. Additional examples are provided in Table 1.4. The emerging gain vs. cost of collaboration trade-offs among the participating entities can be identified even for such networks that emerged naturally through spontaneous evolution and continue to evolve in most cases. In many cases, it seems that Natural networks form as consequences of serving some gain of collaboration objectives among cells, proteins, etc., and usually once their cost becomes greater, e.g., as cells become old, vessels and receptors age, etc., the networks progressively

---

[4]Malware is a generic term to denote spreading malicious software cumulatively.

become dysfunctional and eventually cease their operation, fulfilling their scope.

The second broad category of complex networks includes those that were initiated by humans, but their operation and evolution is not controlled by humans, at least not currently or in the near future. In fact, one of the underlying goals of this book is to set a framework for enabling analysis and control of networks of this category, and this within potentially multiple application frameworks. Characteristic examples of this complex network category are networks emerging in finance, information, and news dissemination networks, rumor spreading networks, malware propagation networks, production and supply chain networks, etc. More extensive examples are provided in Table 1.4. For this category of complex networks, the benefit part of the formation trade-off is evident. Usually such networks are created in order to serve a practical need, or operation of daily human life. However, this comes at an operational cost, which can vary in form and performance. The objective here is to properly control this trade-off in order to increase the gain with the minimum possible cost. However, sometimes it might be required to sacrifice some of the benefit in order to obtain much larger cost savings (and vice versa when the trade-off is balanced towards the cost minimization side).

The third class in this classification is the completely engineered complex networks. Such networks have been conceptualized and created artificially, and most of the time we have control over them, at least up to an acceptable degree. Communications networks are a typical example. They were humanly designed from their inception, then reached full-scale, covering the globe, and nowadays administrators and investors control the way they expand and evolve. Similarly, transportation networks and their subnetworks, such as air-traffic, roadmap, and sealine networks, have been developed and are constantly adapted to fit the needs of humans regarding traveling and product transport (logistics). More examples of such networks are provided in Table 1.4, along with application perspectives of their operation. The gain vs. cost of collaboration of the network entities has the same form as for human initiated/spontaneous evolution complex networks. Such networks have been designed to offer some benefit, e.g., data bits transferred in communications networks, while incurring some operational cost, e.g., energy in wireless networks. Balancing such trade-offs is the main objective of engineers, and usually adapting it to operational requirements or society trends becomes another important objective.

Another important network category, and perhaps the most useful from the perspective of complex network analysis classification, is the one more related to the structural nature of each complex network type and it is mainly based on the mathematical representation of such networks as graphs. By structural nature, we refer to the interactions developed between the entities of a network and the properties/features of the entities and their interactions cumulatively. This mathematical representation will be the explicit topic of the next chapter (Chapter 2) and the specific features of each separate category of network classes according to their structure will be provided in later chapters (Chapter 5 and Chapter 6). Here we only provide the corresponding classification with some characteristic examples of complex networks, and postpone analysis for the following chapters. The list of examples is non-exhaustive and mainly aims at providing an overview of the corresponding network instances, based on which the corresponding network types can be identified in practical cases. Table 1.5 contains the classification of complex networks according to their underlying structure. This could be considered the most important classification of complex network from an engineering and scientific point of view, since it is indicative of the expected properties and behaviors developed or emerging in each type of network. We should note that a scale-free network mentioned in the table is essentially a network whose degree distribution follows a power-law, at least asymptotically. Thus, for the rest of the book, we will employ the term "power-law" to denote networks following exactly a power-law degree distribution and "scale-free" to denote those that follow such degree distribution asymptotically.

In many cases, it suffices to accurately identify the type of a complex network, and then employ standard methodologies developed for the different classes of network types. It should be explicitly noted that a network is always a representation or model of observable reality, but not the reality itself. As a representation model, the network does not always provide the complete information associated with its actual representation. However, it does explain the basic mechanisms and the characteristic functions/features the network has compared to other types of networks. The latter must be done in a manner where no two networks having different operations overlap in terms of modeled behaviors by the corresponding representations. Thus, the representation should bear the properties of uniqueness and accuracy, in order to be able to distinguish between different types of complex networks and at the same time be able to use them properly for the analysis and control of their functions. A third element for each representation is efficiency/convenience. This means that a representation should be manageable in terms of space and complexity requirements for the current technological potentials available. A representation requiring significant amounts of storage memory or one that cannot be processed in the amount of time allowed by the corresponding application framework is unsuitable and essentially of no use. Efficiency in storage and handling convenience in the manipulation and exploitation are essential elements for a network representation as well.

Among the types of complex networks provided in Table 1.5, this book will focus strongly on complex communications networks and especially wireless networks. Wireless distributed devices have nowadays dominated their desktop counterparts and it is expected that soon the wired infrastructure will mainly be restricted to a backbone carrier role, while the wireless will not be just a plain access last-hop interface, but rather an added-value flexible and autonomous network that uses the wired core for overseas and long-distance

Table 1.5: Topology-based complex network classification.

Network type	Network examples
Regular	Lattices Grids Crystals Chains Optical ring networks Cellular phones Supercomputing infrastructures Cloud services Sensor
Random	Peer-to-peer Gas molecules (in equilibrium) Brownian motion email virus netoworks Grid percolation Immunization networks
Mesh	Sensor Delay-tolerant networks Optical networks ZigBee/Bluetooth LTE-A (4G) WiFi (802.11x networks)
Power-law	Metabolic Population of cities Word frequencies Co-authorship networks Affiliation networks Neurons
Scale-free	Social networks WWW Internet (AS[a] routers) DNS[b] routers Protein interaction networks Inter-bank payments Airline networks
Multi-hop	Military networks TETRA Packet radio networks (CSMA/CA) Sensor Vehicular Roadmaps LTE-A (4G) networks Cognitive Radio networks

[a]AS: autonomous systems.
[b]DNS: Domain Name Service.

connections only [75]. It is envisioned that the bulk of the local traffic will be transferred through wireless channels, and in addition, enhanced and demanding applications will be locally supported by wireless networks in order to decongest the wired core [111].

On the other hand, wireless distributed networks, such as ad hoc, sensor, mesh, vehicular, delay tolerant and WiFi, exhibit several impairments and do not inherently support a seamless transition from a wired-oriented application design philosophy widely followed until today to a wireless-centered data transmission expected in the future. In addition, the underlying technology has not exhibited the respective development of the application layer, e.g., the tremendous proliferation of online social networks. The latter inherently bear some features, such as logarithmic scaling and robustness, that would mostly be desired in artificial networks. For this reason, it is strongly required that wireless distributed infrastructures are modified in a fashion that ensures the realization of the demanding and diverse applications, without significantly impacting their operation and resource management. Network modifications should be as transparent as possible and they should allow the maximum possible flexibility between the original and induced networks, enabling on-demand responses to different operational time scales and requirements. This would also align the progress in content and infrastructure, thus smoothing out the transition from legacy systems, where the main information sharing paradigm is that the user acts solely as information consumer of the offered services and data, to more advanced and efficient architectures, where the end-user has turned into both information consumer and producer [122], given the capabilities offered by the current consumer electronic devices.

In the next section, complex network analysis (CNA) will be put into a broader perspective for Network Science research and the corresponding content, status, and challenges faced by network engineers. It will essentially provide a broader perspective for the methods and approaches presented in this book.

1.3   Network Science

The previous discussion on complex networks represents a cumulative effort to develop models of emerging network structures, irrespective of their potential application framework and any practical use they might have. However, as in all other scientific fields, for complex networks too, it is desired to develop a broader framework, where it will be possible to combine theory with application and create a strong bond between modeling and practice.

Network Science is a term that has been coined lately to denote exactly this broader and more ambitious effort for a proper scientific field devoted solely to the study, analysis, and applications of networks, wherever and whenever they emerge. A simple definition of Network Science is the following:

Definition 3 *Network Science is the organized knowledge of networks based on their study and by using formal scientific methods.*

This definition should be used with caution, since it explicitly segregates the scientific from the technological part of network study. However, it should be noted that within the Network Science framework, building and extending the scientific part takes place with the application perspective in mind, and oftentimes, as it will be explained and shown in the sequel, the application drives the theory as well.

Focusing on the case of information networks, which will be the main topic of this book, the components of modern communication and information networks are the result of technologies, which are based on fundamental knowledge emanating from physics, mathematics, circuits, systems, and even material science in various capacities. For instance, computer networks were developed due to advances in circuits and signal processing, while wireless communications networks were enabled due to advances in physics, materials, and computer science. Especially for these types of networks, namely communications, several advances and novelties of smaller scale and in various other areas as well have enabled the design of modules critical for their operation/performance. The assembly of all such novelties into the development of networks, however, is based extensively on empirical knowledge rather than on a deep understanding of the principles of network behaviors gained from and underlying the science of networks. For example, regarding the emergence of the protocol layering concept and the infamous TCP/IP protocol stack used extensively in modern networks, the technology was first developed within the industry and it is only lately that it was possible to consider the whole of the protocol stack across layers through a uniform mathematical methodology (Network Utility Maximizaion—NUM), and thus, essentially reverse-engineer the whole stack [46]. Now it is possible to optimally select the parameters of various mechanisms across layers through NUM, where the design choices will be holistically optimal. However, this was not possible a few years ago for the aforementioned reasons. Another similar example is metabolic networks, for which the consideration of network theory has only lately proved to be fruitful and employed more systematically.

Considering the field from a holistic perspective, it becomes evident that practitioners in each major application area of Network Science have their own local nomenclatures to describe network models of the phenomena in which they are interested and their own notions of the content of Network Science. For example, spatially distributed networks emerge both in communications networks and topography (multi-hop and roadmap networks respectively), or in biology and computer networks (small-world and scale-free). However, different terms have been traditionally employed for the same concepts and slightly varying mechanisms employed to solve the same problems, not to mention the same computerized tools employed. Consequently, a new field of network investigation is yet to be codified and shaped properly, in order to reach the maturity level of the rest of the scientific fields, like fluid mechanics, materials, etc.

The field of Network Science is currently evolving in a lively way and there is limited concrete understanding of its ultimate scope and content. However, as so many other scientific disciplines in the past, Network Science will eventually evolve into whatever its practitioners create in the coming time period. Similar examples are the greater fields of probability and random processes, fluid dynamics, etc., where applications based on empirical knowledge were massively employed before the knowledge that is available today was established. However, it was exactly the formal shaping of these fields that helped in better understanding the underlying phenomena, achieving more progress in the theoretical part, and eventually obtaining desired and efficient control of the applications as well.

There has been no complete theory until today offering the required fundamental knowledge for analyzing and designing large (or arbitrary scale) complex networks in a way that network or network component behaviors can be predicted prior to realizing them in everyday life. This is especially true for human initiated networks with spontaneous evolution and of course natural networks. Even most of the technological networks designed, operated, and controlled by humans would need to bear such features more often and more efficiently. Admittedly, Network Science is one of those fields where technology evolved much earlier than scientific knowledge. In fact, it is now widely accepted that the main highlight of Network Science is unfortunately that currently it essentially describes a field of fragmented research. Network Science consists of the study of network representations of physical, logical, and social interactions, leading to predictive models of these phenomena and relations. The fragmentation of knowledge is meant in the sense that most of the methods and approaches employed for the study have a specific character and they are applied in a narrow-minded fashion within each discipline span. Thus, even though a method developed could be properly adapted and employed in more than one discipline, e.g., biology and communications, it is only lately that this possibility and its benefits have been identified and appreciated. In addition, the current knowledge about the structure, dynamics, and behaviors of both large infrastructure networks and vital social networks at all scales is primitive [117]. This fragmentation is largely evident in that not all disciplines of Network Science scale at the same order, namely, some are progressing well, e.g., communications networks, while others at a much slower pace, e.g., financial networks. Additionally, it has not been possible to perform cross-disciplinary evaluations to the desired degree, due to such fragmentation.

The communities from which Network Science is expected to emerge encompass many different and diverse disciplines of applications areas. Characteristic examples are, among others, the biology field, which seems to provide the most diverse examples of observed complex network structures of arbitrary order and capabilities, all working usually in conjunction and in an efficient manner. Telecommunications networks have employed substantial mathematical tools for their analysis and development and could provide solid background for quantitative methods in many other disciplines in Network Science and complex networks. Furthermore, online social networks and communities could provide flexible and controllable ground for emulating and evaluating various mechanisms of broader interest and thus may constitute a realistic testbed infrastructure for mechanisms of other disciplines. Other disciplines expected to contribute in the development of Network Science include mathematics and particle physics, statistics, etc.

By considering all these facts, there seems to be a widespread realization that codifying a common nomenclature and body of core knowledge is at least useful, if not inevitable, in order to achieve the level of control and efficiency desired. Even though this has not yet occurred, significant progress towards that scope is accumulating within the Network Science framework, as will be explained in more detail in the forthcoming subsections.

1.3.1 Content and Promise of Network Science

As explained already in the previous subsection, Network Science aspires to become a broader field of science, in which the theoretical and technological knowledge related to emerging networks in multiple and diverse facets and ways will converge. As a prominent field of its own merit, the content of Network Science can be implicitly defined through the set of encompassing core principles, broadly embedded in quantitative approaches. Such quantitative disciplines are meant in the systematic manner that they could be potentially taught to students, in order to prepare a newer generation of network expert practitioners, researchers, and professionals, as with other established disciplines.

However, passing the torch to a more conscious and better-trained generation means having an already concretely established framework and future horizon for emerging problems and desired achievements. Although the boundaries of Network Science remain fuzzy, as was the case with many other currently well-defined scientific disciplines, there is broad agreement on key topics that should constitute the field, the types of tools that must be developed and employed, and the research challenges that should be explored. Today, there is a consensus among the practitioners of research on networks for physical, biological, social, and information applications on the topics that constitute Network Science.

One of the initial questions that emerged in the early days of Network Science was the extent to which current research on networks exhibits a core content that cuts across the involved diverse applications areas of interest. Various studies and observations have revealed common elements in diverse applications that helped in creating an operational definition of the field pertinent to the task of the field, as has been conceptualized distributively by many already active researchers and professionals, originally coming from diverse disciplines and bringing in different methodologies.

The core content of Network Science is basic scientific approaches, currently consisting of simplified models and of techniques that are appropriate for the analysis of small-world networks that exhibit low topological complexity. Such networks mainly include the earlier communications networks, some online social networks, and some other engineered networks. Also, in some other large-scale and highly complex networks, such as the human nerve system, there are several subsystems consisting of smaller networks, which fall under the previous umbrella, and for which acceptable models have already been developed, paving the way for more advanced and accurate models.

This common core of Network Science is the study of complex systems (networks) whose behavior and responses are determined by exchanges and interactions between subsystems across a well-defined (possibly dynamic) set of pathways. The central point is that the behavior of a network is determined both by the pathways (structure) and by exchanges and interactions (dynamics). Moreover the structure itself may be (and usually is) dynamic. The analysis of network structure is currently more advanced than that of network dynamics. The outputs of models analysis in the core content are insight and quantitative understanding, not engineering design. The objective of this field is to shift gears and enable the convergence of established knowledge in structure and dynamics to an equal degree.

It was already explained that current research on networks is highly fragmented and usually conducted in disciplinary settings. However, as for complex networks, the field of Network Science is vast and inherently multidisciplinary. Network structures seem to emerge naturally, essentially whenever information exchange or control signalling emerges. And following the emergence of network structure, dynamic network behavior, mainly involving interactions, structure evolution, control, and computation, emerges as well in multiple ways. For instance, such dynamic behavior could be a highly complex structure modification imposed by network factors and determined and executed in an autonomic fashion, as it is the case in several biological networks, or a simple data packet forwarding which takes place in high speed switches. Numerous intermediate combinations of complex network structures, behaviors, and dynamics exist in between these cases and in all such examples, elements from different and diverse disciplines emerge. For instance, even though structures can be the same for different disciplines the cumulative observed behaviors can be considerably different and vice-versa. Such diversity increases the complexity of analysis, but also allows for more flexibility and potential benefit, once such diversity is mastered and exploited for useful purposes.

If Network Science is to eventually exist in a meaningful way, the approaches developed and used within it must also be effective over many application domains, with well-understood techniques to apply general tools, methods, and models to specific domains. This book will be involved in various capacities with all such considerations and potentials of Network Science, as will become evident in the rest of the chapters.

1.3.2 Networks and Network Research in the 21st Century

The content of Network Science has been extensively studied in the previous section. However, in order to understand completely the field and its actual content, it is proper to study in parallel a few basic moments from the evolution of the emergence and study of networks.

Network Science is a term that emerged gradually through the concept of network-centric warfare developed by the US Army. It is no coincidence that the first data networks also developed under the same regime. As mentioned before, the term has been used lately to cumulatively describe the research centered around the study of networks emerging in any possible discipline and application. Nevertheless, the emergence of networks is much older, even older than the human species.

The first networking structures, which were used to transfer chemical messages, were developed in the primitive life forms. Similar messaging pathways (nowadays represented by networks) developed in chemical bonds in physical elements and substances from the very early days of the Earth and long before human species emerged on the planet.

On the other hand, humans are social entities and networks have formed by their interactions in various capacities from their very early emergence. Their interactions have lead to various developments, such as open markets, global economies, etc. Financial networks have emerged in these as well, following human operation, behavior, and trends. And even though financial networks have emerged from the early days of currency markets and have proliferated following the market and trade proliferation, the first networks to have been systematically studied are the computer and later communications networks.

In addition, computer networks are the first networks that were first analytically designed/studied and then actually implemented. Their evolution followed the knowledge accumulated their analytical design and ever since has followed a development loop, where progress is dictated either from technology or theory, depending on the cycle phase, so that once technology develops due to deeper knowledge it immediately spurs a frantic research, which in turn leads technology to even higher complexities and benefits. Communications networks have been tightly related to the US Army research. As already stated, the first data network, ARPANET, was build with funding by DARPA, in order to interconnect and promote research among American universities and research centers, some of which belonged to the Army.

This observation has increased the interest for quantitative and qualitative study of networks, calling for a holistic analytical science of networks, namely Network Science. Such interest and essentially various requirements posed by the current trends of societies call for more holistic development tools, such as the one described above for computer/communications networks and more inter-disciplinary outcomes, as explained before.

Human understanding of networks has the potential to play a vital role in the 21st century, which is witnessing the rise of the Age of Connectivity. There seems to be an enormous demand for information on how to design and operate large global networks in a robust, stable, and secure fashion, following the tremendous generation of information in a previously known Information Era. If information has been the power of the current century and it is going to play an equally significant role, if not even more significant, in the future, the means to exchange information, process it efficiently and distributively when local means are not sufficient, and share the obtained results will become even more important. The prospect of the content and results of Network Science in this framework become even more significant than before.

Present systems used by governments and other services need to be improved and integrated into a solution encompassing the physical, cognitive, and social domains, as well as the information domain. This is an elusive objective that involves both near- and far-term efforts in network research, and includes efforts that go beyond research per se.

Network Science consists of multiple branches as diverse as the disciplines that shape it. The practitioners and researchers involved in one subset of them usually find it easy to understand problems and notions in others as well. In addition, a recent trend is that many experts from one discipline are becoming involved in some other branch of Network Science, in order to build a relevant background into methods of relating disciplines and thus obtain new perspectives in the understanding of their own fields of expertise. Cross-disciplinary research involving various Network Science branches seemingly different is also becoming a norm and many research centers are adopting such approaches, bringing together dedicated diverse teams that study the computational and experimental aspects of social, biological, and communications networks, and their ties with networks from other disciplines.

In the following, we provide a brief list of some of the most important branches of Network Science with inter-disciplinary interest, as they have emerged in the literature and relevant fora until now. Most of them contain cutting-edge mathematical methodologies and numerous open problems, some of which appear to be considerably tough:

- modeling, simulation, testing, and prototyping of very large networks

- command and control of joint/combined networked groups

- impact of network structure on organizational behavior

- security and information assurance of networks

- relationship of network structure to scalability and reliability

- managing network complexity

- improving shared situational awareness of networked elements

- enhanced network-centric mission effectiveness

- advanced network-based sensor fusion

- hunter-prey relationships

- swarming behavior

- metabolic and gene expression networks

- visualization tools

Network research in general has proven to be an important source of economic growth via the creation of new commercial pathways not seemingly evident before. In fact, the very recent network research is leading to new and growing businesses like Internet and large data analytics, social media, etc. One of the outcomes and benefits of the above thinking regarding network research is the emergence of various data mining and exploiting systems/engines, which emerged lately and generated not only significant economic growth, but also contributed to a paradigm shift of the Web and the way we treat data information anywhere it comes from.

Although the technology for constructing and operating engineered physical networks is sophisticated, critical questions about their robustness, stability, scaling, and performance still cannot be answered with confidence without extensive simulation and testing. Investment in Network Science is both a strategic and urgent national priority for some countries, most prominently the United States [117]. This book aspires to further contribute towards this direction.

1.3.3 Status and Challenges of Network Science

Based on the above discussions it becomes evident that the current status in the field of Network Science has not converged to a stable direction with well defined boundaries, as is the case for other well-established scientific branches.

Network Science is a field of research and practice that emerged in the last decade and its whole boundaries are still evolving. New application domains are discovered and new opportunities for cross-disciplinary research are identified. In addition, currently there exists a gap between available knowledge about networks and the knowledge required to characterize, design, and operate the complex global, physical, information, biological, and social networks on which the well-being of mankind has come to depend [117]. Just like the radar technology awaited the basic science of electromagnetism and that of mobile communications the development of radio science and circuits, the ability to control the complex networks in our lives awaits as yet unforseen discoveries in the science of networks.

Presently, the Network Science community is a worldwide and diverse research community with shared concepts and concerns. It has been found that even within diverse subgroups of this broad research community, consistent and convergent opinions exist. A significant percent of them (slightly a bit more than 2/3) agree on the fact that Network Science is a definable field of investigation, even though not yet absolutely and accurately shaped.

In general, three main dimensions account for the difficulties emerging in the research challenges faced within the framework of Network Science and relevant applications. These "umbrella-like" areas, as can be identified in the literature, are summarized as follows:

- complexity

- wide range of interacting scales

- network-to-network interactions

The first, cumulatively contains emerging issues that have to do with computation, communication, and interaction between network entities. Complexity may characterize the encountered difficulties for processing or exchanging information in massive fashion, or it could describe the difficulties emerging in handling various representations of the networks. The second dimension contains the wide range of interacting scales of the various networks in the framework of Network Science. In a system, different types of networks may interact with each other. However, such networks could possibly have different scaling behaviors, in terms of size, timing, etc. Finally, the third dimension involves the types of interactions between different types of networks, e.g., cyber-physical systems [108], [105], but also interactions between nodes of the same or different networks (in terms of scale and type). All the above dimensions are quite broad and frequently overlap, posing even more complex challenges for the researchers involved. In the sequel, we will provide some more specific challenges emerging in the field of Network Science, in order to provide a more concrete picture of the current status and scope of Network Science.

Given the latest updates and the current status in Network Science one may identify seven major classes of challenges, which attract the interest of the scientific and industrial communities and will have a significant role in the evolution of the field. These directions have broadly appeared in the literature and we summarize them in the following list:

- Dynamics, spatial location, and information propagation in networks

- Modeling and analysis of very large networks

- Design and synthesis of networks

- Abstracting common concepts across fields

- Better experiments and measurements of network structure

- Robustness and security of networks

- Increasing the level of rigor and mathematical structure

The first one is about emerging problems related to the dynamical and topological behavior of distributed structures, such as ad hoc networks in complex communications networks, and the transfer/distribution of information through them. The second is purely of a topological nature and involves the modeling and analysis of very large networks, in terms of optimal topologies, as well as the means to represent and handle such structures. The third is about the very existence of networks, namely the ability to design and synthesize the desired networking structures in various disciplines and especially for critical applications. The fourth regards the establishment and expansion of cross-discipline research, where common concepts among Network Science disciplines are abstracted and exploited across fields, by developing generic methodologies, and tools. Especially towards the latter direction, it will be required to progress both in our experimentation and testing means, in order to be able to collect more accurate measurements and data, on the behavior and performance of the studied networks and developed methodologies respectively. This is the topic of the fifth research direction, while the sixth involves a very popular concern spanning our societies, namely security and robustness. The latter regards the sustainability of the networks that are of significant interest or play a critical role, while the first is about the emerging behaviors and outcomes of these networks. Finally, perhaps the most important challenge will be to increase the current mathematical rigor, thus providing more flexible, accurate, and sophisticated means to achieve all the previously mentioned challenges.

It is expected that connections between the basic and applied portions of the research will be much more intimate. One of the envisioned scenarios is the application of modern communications networks and tools and the insights of modern social network theory to transforming the management of educational projects, and possibly other research stimulating efforts.

Since Network Science is at an early stage of its development, a broad portfolio of basic and applied research is expected to create greater value than a more focused portfolio. For this reason, this book will initially cover the broader span of Network Science and then take this one step ahead by presenting a more focused application domain and corresponding methods. Research on the lower layers of the network architecture is relatively mature. Thus, the most immediate payoffs from Network Science are likely to result from research associated with the upper levels of the network architecture and the social networks that are built at an even higher level upon their outputs. This aspect will be covered by the later chapters of the book, as it will become more specific.

Chapter 2

Basic Network Graph Models and Their Properties

Towards building the hierarchical approach of this book for the study of evolutionary dynamics, the first step is to study the basic network models and properties established, and then use them constructively for more advanced studies. The examples provided in the previous chapters of various network types, operations, behaviors, mechanisms, and emerging behaviors in Network Science in general, have motivated the need for a more systematic analytical approach of their analysis and control.

Before we present the more advanced analytical tools and techniques for exploiting the potentials of complex networks and Network Science, in this chapter, we provide a concise summary of the fundamental mathematical notions and techniques of network engineering. These include elements from traditional Graph Theory and the more advanced field of Random Graph Theory, the latter being the first step towards the study of dynamic network behavior. We first present traditional Graph Theory and then separately Random Graph Theory, even though both are general Graph Theory toolboxes, due to the different modeling approach adopted in Random Graphs. Random Graph Theory considers a dynamical network behavior and thus a probabilistic modeling perspective is developed, whereas in traditional Graph Theory deterministic models and properties are studied.

This chapter will provide a summary of the fundamental notions that emerge in the study and analysis of networks, in addition to a common mathematical language that is able to describe accurately the fundamentals of network structure and dynamics. Graph Theory has been and remains the main mathematical tool that ensures accuracy and control. It is also the main language for communications networks, which is the main network type of interest in this manuscript among those considered Network Science. A fair portion of this chapter will be devoted to the introduction of these concepts through a concise summary of the basic concepts, measures, and properties of the most important fields of traditional Graph Theory. This summary will constitute a ground-breaking building block for studying networks from different disciplines in a generic fashion and will enable further extending this toolbox for the analysis and exploitation of more advanced concepts in Network Science.

Regarding Graph Theory, the basic definitions and properties of all graphs will be presented, followed by definitions and principles of graphs that have to do with paths and cycles, connectivity, flow through networks, planarity and coloring (covering) properties. An alternative, but very popular algebraic representation will be also presented, since in modern complex network analysis it seems that this graph model representation and analysis approach is gaining ground compared to the traditional ones.

Furthermore, Random Graph Theory is a combination of concepts from traditional Graph Theory and Probability Theory and has been proved invaluable in the study of the dynamic, evolutionary behavior of several simple network problems. It emerges as the most appropriate technique for more advanced studies of dynamic network evolution. The various Random Graph models and their equivalences will be presented, followed by properties of all graphs and the emergence of threshold behaviors, which seems to be the norm in relevant observed phenomena.

At the end of this chapter, we provide a concise summary of notation of the most basic elements emerging in Graph Theory. Such notation will be extensively used in the following chapters. Though Graph Theory notation is non-standard, the aim of this table is to enable the reader in his/her study, by providing a quick reference when necessary. In most other cases, the notation employed in the following chapters remains self-explanatory.

2.1 Graph Theory Fundamentals

As already explained, the diversity of the field of network science calls for methodologies and frameworks where the study of emerging network structures and behaviors will be studied in a generic fashion. Graph Theory provides a mathematical language for properly describing the interactions between various agents in the considered networks. The value of the means provided by Graph Theory is significant, because the quantitative tools developed are independent of the involved application framework. In fact, the mathematical representation of Graph Theory enables suppressing the effect of discipline-specific parameters and revealing the actual dynamics of the studied network and identifying parameters enabling their control, irrespective of the scope of a specific network. Graph Theory is the fundamental and essential language for basic and advanced network analysis and engineering.

The notion of graph is extensively used to represent various types of networks. A graph is an ordered pair $G = (V, E)$ where V is the set of vertices of the graph with cardinality $|V|$ and E is the set of edges of the graph with cardinality $|E|$. The edges of a graph are two element subsets of V. In this book, unless otherwise stated, we represent an edge between two vertices i, j with (i, j) as shown in Figure 2.1.[1] There are many categories of graphs with respect to the developing interconnections between agents and emerging behaviors. In the previous chapter several classifications of these important network types for Network Science and complex communication network design have been presented in more detail. In this chapter, we will assume a simple and generic model of graph, as will be described in the following subsections, and present properties that hold in general for many types of graphs. Additional properties, features and characteristics that apply for each specific type of network will be provided and analyzed in the following chapters. Progressively more advanced concepts on networks and features will be presented, analyzed, and prescribed.

At this point we should note that despite the fact that Graph Theory now has a well established history in mathematics and other scientific areas, notation is not yet completely standardized. This is partly due to the diversity of disciplines employing it and contributing to it. In this book we employ the most widely accepted nomenclature for the most important or required elements. We also provide a summary table in the end of this chapter (Section 2.3) with all Graph Theory notation used. In any case, most of the quantities required for the study of complex networks are self-explanatory and symbols will be compatible with them.

In the sequel, apart from basic definitions, we often include several useful properties of general interest (usually in the form of theorems) for the analyzed structures and parameters. Unless otherwise noted, these constitute well-known results, which we include for completeness purposes and in order to provide a quick reference for the interested researcher or student. We provide them without proofs and accompany them with suitable references for the more interested reader, who wishes to spend more time focusing on the details of the provided methods.

2.1.1   Basic Definitions and Notation

In this subsection, we provide the most basic definitions regarding graphs and networks, along with their notation and simple emerging properties. More advanced concepts are provided in the following subsections.

An *undirected graph* $G(V, E)$ consists of a pair of finite and nonempty set $V = V(G)$ of $|V| = n$ points with a set of $|E| = m$ unordered pairs $(i, j), i, j \in V$ of distinct points of V. Elements of V are referred to as points or

---

[1]Typically ij denotes an unordered edge between vertices i, j, while (i, j) denotes a directed edge from node i to j. When there is no risk of ambiguity, the (i, j) notation is employed for both cases.

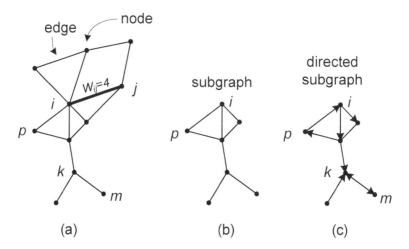

Figure 2.1: Fundamental definitions on deterministic graphs.

vertices and elements of E are referred to as lines or edges or links (Figure 2.1). All these terms are used interchangeably, e.g., points-nodes and lines-edges-links, respectively. Sometimes an edge $x = (i,j)$ may be denoted as $x = ij$. We say that i and j are adjacent vertices, vertex i and edge x are incident with each other, while two distinct edges x and y incident with a common vertex are called adjacent edges. Sometimes vertices i, j of line x are called endpoints of x as well. These definitions are illustrated in Figure 2.1.

A directed graph or digraph consists of directed edges and therefore the pair (i,j) is ordered, expressing an arrow beginning from node i and pointing to node j, as noted above. In the directed case, the edges (i,j) and (j,i) are different (which is not the case for an undirected graph) and the existence of one of these does not necessarily imply the existence of the other.

A graph $G' = (V', E')$ is a *subgraph* of $G(V, E)$ if $V' \subset V$ and $E' \subset E$, where E' is defined on V', and this is denoted by $G' \subset G$. Figure 2.2(a) shows an original undirected graph G and Figure 2.2(b) a subgraph G' of G. If G' contains all edges of G that join two vertices in V', then G' is called the *induced* subgraph or the subgraph *spanned by* V' and it is denoted by $G[V']$ (Figure 2.2(c)). If $V' = V$, then G' is said to be a *spanning* subgraph of G (Figure 2.2(d)).

If $W \subset V(G)$ for a graph G, then $G - W = G[V \backslash W]$ is the subgraph of G obtained by deleting the vertices in W and all edges incident with them. Similarly, if $E' \subset E(G)$, then $G - E' = (V(G), E(G) \backslash E')$. If $W = \{x\}$ and $E' = \{xy\}$, then the above notation is simplified to $G - x$ and $G - xy$. Finally, if x and y are nonadjacent vertices of G, then $G + xy$ is obtained from G by joining x to y. The complement $\bar{G}$ of $G = (V, E)$ is defined as $\bar{G} = (V, V^2 - E)$.

The *order* $|G|$ of G is the number of vertices $|G| = |V(G)|$ and the *size* of G, denoted by $e(G)$ is the number of edges in G, i.e., $e(G) = |E(G)|$. Both the

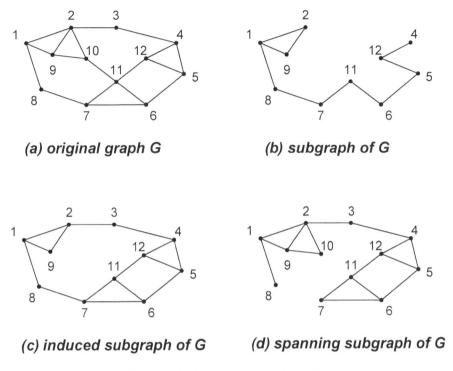

(a) original graph G

(b) subgraph of G

(c) induced subgraph of G

(d) spanning subgraph of G

Figure 2.2: Definitions of subgraphs.

order and size parameters of a graph are characteristic of the scales that the network follows, in terms of the entities involved in the network and/or their interactions, respectively. Usually, the scaling behavior as these parameters increase is of great interest, indicating the potentials and usefulness of each specific network graph type.

The neighborhood of a node $x \in G$, denoted by $\Gamma(x)$ or most commonly in communications by $\mathcal{N}(x)$, is the set of all vertices of G adjacent to x. The degree of a vertex i, $d(i)$ (usually denoted by k_i to avoid misunderstanding with distance functions) in an undirected graph is the number of edges having as one of their endpoints the vertex i. In directed graphs, each vertex is characterized by two degrees, the in-degree k_i^{in}, which counts all edges pointing to node i and the out-degree k_i^{out} counting all vertices starting from node i (Figure 2.3). In both the directed and the undirected case, we denote by $A = [a_{ij}]$ the adjacency matrix, where $a_{ij} = 1$ if there is a link from i to j, otherwise $a_{ij} = 0$. By using the adjacency matrix, we obtain $k_i^{out} = \sum_{j=1}^{N} a_{ij}$ and $k_i^{in} = \sum_{j=1}^{N} a_{ji}$. A is symmetric if it refers to the undirected case. More on the adjacency matrix and other matrices that characterize different graphs will be provided in Subsection 2.1.8, which focuses on spectral Graph Theory. It will also be shown how the algebraic (matrix) representations of networks

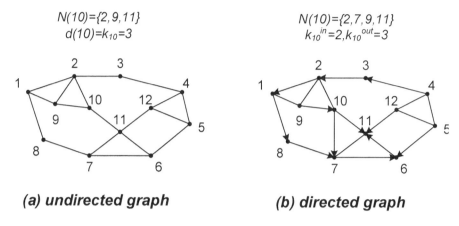

(a) undirected graph   **(b) directed graph**

Figure 2.3: Examples of neighborhoods and node degrees.

are capable of providing efficient tools not only for the analysis of traditional network problems, but more importantly, that they constitute tools for the more interesting analysis of the dynamic behavior and/or evolution of networks.

Similarly to adjacency matrices, the *incidence matrix* of a graph G may be defined. If G is a bidirectional graph, then the incidence matrix $B = [b_{ij}]$ is a $|V| \times |E|$ matrix such that $b_{ij} = 1$ if vertex v_i is incident with edge x_j and zero otherwise. For the case of a directed graph G, the incidence matrix is again a $|V| \times |E|$ matrix, such that $b_{ij} = -1$ if edge x_j leaves vertex v_i, $b_{ij} = 1$ if edge x_j enters vertex v_i and zero otherwise.

A *clique* in an undirected graph is a subset of its vertices such that every two vertices in the subset are connected by an edge. They are rather important concepts in Graph Theory and Network Science in general, and several problems reduce to finding cliques in graphs representing network topologies. A *maximal clique* is a clique that cannot be extended by including one more adjacent vertex, that is, a clique that does not exist exclusively within the vertex set of a larger clique. A *maximum clique* is a clique of the largest possible size in a given graph. The clique number $\omega(G)$ of a graph $G(V,E)$ is the number of vertices in a maximum clique in G.

A *d-regular graph* is an undirected graph where each vertex has the same degree equal to d. A *complete graph* is an undirected graph where all vertices are connected with all other vertices and thus, it is an $(n\text{-}1)$-regular graph. Regular networks emerge oftentimes in nature and occasionally in engineered applications as well, such as communications and power networks. These structures will be extensively used in various capacities in the following chapters of this book, mainly as reference network models or as starting points for obtaining through evolutionary processes other network paradigms.

Definition 4 *Given a graph G, its corresponding line graph, $L(G)$, is a graph* *constructed as follows:*

- *Each vertex of $L(G)$ represents an edge of G.*

- *Two vertices of $L(G)$ are adjacent if and only if their corresponding edges are adjacent in G.*

Definition 5 *A graph (directed or undirected) is weighted, if a measurable quantity (referred to as weight and usually denoted by w) is assigned to each edge, $w : E \rightarrow \mathbb{R}$.*

Such quantities-weights might represent, for example, costs, lengths, or capacities in communications networks, distances, interest in social networks, cash flow in financial networks, and numerous other context values depending on the specifics of each problem. Similarly, with the adjacency matrix, the weight matrix $W = [w_{ij}]$ can be defined, where w_{ij} is the weight of the link (i, j). Furthermore, especially for weighted graphs, except for the degree of a vertex (or in-degree and out-degree in digraphs) which does not take into account at all the weights of links, a joint metric of both node degree and adjacent link weights might be defined, denoted by strength of each node (or the in-strength s_i^{in} and the out-strength s_i^{out} correspondingly for digraphs). Node strength expresses the total amount of weight that reaches or leaves node i correspondingly. Thus,

$$s_i^{out} = \sum_{j=1}^{N} w_{ij} \text{ and } s_i^{in} = \sum_{j=1}^{N} w_{ji} \tag{2.1}$$

In the case of undirected graph-network case, $s_i^{out} = s_i^{in} = s_i$ where s_i is the strength of node i

$$s_i = \sum_{j=1}^{N} w_{ij} \tag{2.2}$$

2.1.2 Additional Definitions

The previous subsection presented a series of basic definitions regarding Graph Theory, based on which more advanced concepts can be studied and properties of networks analyzed. In this subsection, we provide some more advanced definitions and properties of deterministic graphs, which in turn enable the analysis of frequently emerging behaviors and relations in networks.

Two graphs G and H are *isomorphic* (denoted by $G \cong H$ or more simply $G = H$) if there exists a one-to-one correspondence between their vertex sets, which preserves adjacency. Intuitively, isomorphic graphs describe the same set of relations developing between the underlying players/actors/nodes, i.e., the same networking behavior. Mathematically, graph isomorphism is an equivalence relation on graphs, which means that studying the properties of a graph is equivalent to proving the properties of all graphs belonging to the specific isomorphy class. Thus, from an application perspective, studying one graph from each isomorphism class suffices to obtain a holistic characterization of the specific isomorphism class.

A concept very closely related to graph isomorphism is that of graph invariant. For a graph G, an *invariant* is a number associated with G, which has the same value for any graph isomorphic to G. For example, the order and size of a graph G are invariants. This means that graph invariants are characteristic features of distinct graph types (equivalently isomorphy classes). Moreover, a complete set of invariants determines a graph up to isomorphism. Thus, all isomorphic graphs, namely all those graphs that represent the same type of interactions, bear a specific set of features and properties as well, described by the set of respective graph invariants. Most of the quantities of interest, e.g., the average graph degree, the number of vertices/edges, etc., are graph invariants; nevertheless, not all of them constitute a complete set of invariants. The term "complete set of invariants" characterizes a collection of mappings determining the equivalence of objects in classification problems. In Graph Theory, a complete set of invariants is used to determine an isomorphy equivalence class of graphs. For instance, the order and size of a graph are a complete set of invariants for all graphs with less than four vertices only, which means that using only the invariants of order and size of a graph, all graphs with less than four vertices may be determined.

A *bipartite* graph G is a graph whose vertex set V can be partitioned into two subsets V_1 and V_2, such that every edge of G joins only a node from V_1 with a node from V_2 (Figure 2.4(a)). Such structures where essentially graph vertices are split into two distinct groups and relations take place only between members of different sets emerge often in a special type of social network, i.e., affiliation networks that describe connections between social groups, e.g., advisors-advisees, doctors-patients, etc. A *complete bipartite* graph contains all possible edges joining vertices of V_1 and V_2 (Figure 2.4(b)). A complete bipartite graph with $|V_1| = m$ and $|V_2| = n$ is denoted by $K_{m,n} = K(m,n)$. A *star* graph is a special case of complete $K_{1,n}$ bipartite graph (Figure 2.4(c)). It is also apparent that a $K_{m,n}$ graph has exactly mn edges.

Identifying bipartite graphs is a relatively easy process exploiting the following theorem.

Theorem 1 *A graph is bipartite if and only if (*iff*) it does not contain an odd cycle.*[2]

In the above theorem, the notion of cycle in a graph is exploited. Cycles in graphs are a very important aspect and characteristic feature of them, which will be explained in detail in Subsection 2.1.4. Intuitively, a cycle in an

---

[2]We note that the proof for this theorem, as well as other proofs for the following ones, have been omitted and can be found in various available references in the literature, such as [33], [84], [50].

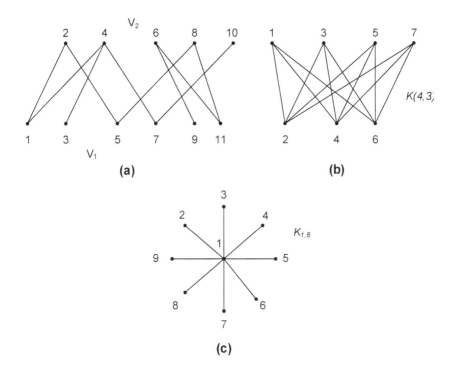

Figure 2.4: Bipartite graphs.

undirected graph is an alternating sequence of vertices and links, which starts and ends at the same vertex.

Thus, in a graph representation, a greedy algorithm deciding on the existence of an odd cycle is sufficient for determining if a graph is bipartite. Of course, complexity and efficiency concerns might call for more advanced processes; however, the above theorem provides a simple and handy characterization for bipartite graphs.

Several special graph types have been identified in Graph Theory. Perhaps one of the most important, both for theoretical as well as application oriented purposes, is that of a tree. Various definitions of a tree graph exist and the following statements are all equivalent for a graph G:

Theorem 2 *The following statements are all equivalent for a graph G:*

1. *G is a tree.*

2. *G is a connected graph and every edge is a bridge.*

3. *G is a maximal acyclic graph; that is G is acyclic and if x and y are nonadjacent vertices of G, then $G + xy$ contains a cycle.*

A spanning tree of a graph G is a tree containing every vertex (node) of G. A *minimum weight spanning tree* (*minimum spanning tree* for short) in a

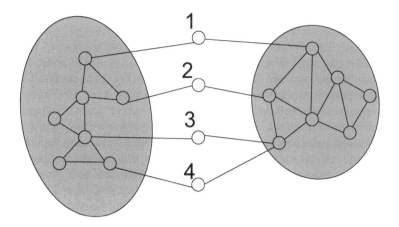

Figure 2.5: Illustration of Theorem 3.

graph. However, at least as many edges as vertices need to be removed, e.g., when the two components of the remaining graph are assumed to coincide, namely the specific vertices are essentially connected to the rest of the graph through a single edge each. In all other cases, more edges will be required to be removed by the definition of edge-connectivity.

One can even extend this result to a more general one, where:

Theorem 4 *For all integers a, b, c such that $0 < a \leq b \leq c$, there exists a graph G with $\kappa(G) = a, \lambda(G) = b$, and $\delta(G) = c$.*

This theorem can be very useful for designing artificial graphs-networks with desired properties on connectivity and edge-connectivity, or redesigning other graphs-networks found in nature so as to obtain more desired connectivity features than those of the original graph-network.

The sparseness or density of a topology, which is reflected in the minimum/maximum degree of a graph G, are also tightly related to connectivity. The following results are characteristic of this relation:

Theorem 5 *If G has p vertices and $\delta(G) \geq [p/2]$, then $\lambda(G) = \delta(G)$.*

and

Theorem 6 *Among all graphs with p vertices and q edges, the maximum connectivity is 0 when $q < p - 1$ and it is $[2q/p]$, when $q \geq p - 1$.*

The first of these two theorems essentially expresses the fact that in a relatively dense graph, where even the minimum degree is connected to at least half of the network nodes, the edge-connectivity is equal to the minimum degree, namely one should remove at least as many as half of the network edges. Conversely, the second theorem says that for a very sparse graph, where the edges are even less than the number of nodes (even less connected than a tree), the graph must be disconnected and $\kappa = 0$.

Definition 8 *A connectivity pair of a graph G is an ordered pair (a, b) of nonnegative integers, such that there is some set of a vertices and b edges whose removal disconnects the graph, and there is no set of $a - 1$ nodes and b edges or of a vertices and $b - 1$ edges with this property.*

It can be easily obtained that for each value of a, $0 \leq a \leq \kappa$, there is a unique connectivity pair (a, b_a). Thus, a graph G has exactly $\kappa + 1$ connectivity pairs.

Results on connectivity may also be extended to cases where all nodes of a network have a minimum of neighbors, namely the minimum node degree information is available. This is reflected in the following definition:

Definition 9 *A graph is n-connected if $\kappa(G) \geq n$ and n-edge-connected if $\lambda(G) \geq n$.*

Connectivity and n-connectivity emerge often in applications of social and sensor networks. In the latter, sensor overlapping is usually required for seamless operation. For this reason, nodes are usually put in grid topologies, where a minimum connectivity is ensured, namely the resulting network is k-connected for some non-zero integer parameter k. Similarly, in social networks, most of the people have at least a minimum number of acquaintances, which also means the corresponding representation graph is n-edge-connected, for some parameter n describing the minimum number of each individual's acquaintances.

A notion closely related to connectivity and searches in evolving networks is that of *commute time*. The commute time $C(u, v)$ between two vertices u and v is the time required to get to node v, starting from node u. It is implicitly assumed that at zero time the visiting process of vertices starts at node u and at each time unit, it transitions to a neighboring vertex according to some policy (in some cases in a completely random fashion, in others according to a more specific policy).

A useful result closely related to k-connectivity, since it involves 3-connected graphs, emerges often in social network applications (see the description of clustering coefficient at Chapter 4, Section 4.4) and it is provided by the following theorem:

Theorem 7 *(Tutte's theorem) A graph G is 3-connected if and only if G is a wheel or can be obtained from a wheel by a sequence of operations of the following two types:*

1. *The addition of a new edge.*

2. *The replacement of a vertex v having degree at least 4 by two adjacent vertices v', v'', such that each vertex formerly joined to v is joined by exactly one of v' and v'' so that in the resulting graph, $d_{v'} \geq 3$ and $d_{v''} \geq 3$.*

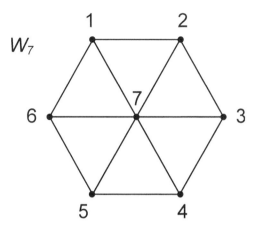

Figure 2.6: Example of a wheel graph.

where for $n \geq 4$, a wheel W_n is a special type of graph, defined to be the graph $K_1 + C_{n-1}$. An example of a W_7 wheel is shown in Figure 2.6.

Furthermore, in many applications or analyses, it is often necessary to split graphs into partially smaller subgraphs, denoted as components. Breaking down a graph into its components, could be imposed by operational objectives and requirements, e.g., splitting communication networks into disjoint routing domains administrated by different entities in order to improve network management, or by analysis objectives, i.e., because in such a way a problem becomes simpler to analyze or handle. For this reason, the following:

Definition 10 *An n-component of a graph G is a maximal n-connected subgraph.*

introduces a concept very often employed in the above settings. A useful result on n-components is described by:

Theorem 8 *Two distinct n-components of a graph G have at most $n-1$ vertices in common.*

Let u and v be two vertices $u \neq v$ of a connected graph G. Two paths joining u and v are called *disjoint* (vertex disjoint) if they have no vertices in common other than u and v. Obviously, this means they have no common edges as well. The paths are *edge-disjoint* if they only have no edges in common. In general, it is intuitive to observe that the connectivity of a graph is related to the number of disjoint paths joining distinct points in the graph. Based on the previous definitions, a set S of vertices, edges, or both vertices and edges separates u and v if u and v are in different components of $G - S$.

Separating vertices and/or edges have significant importance in Graph Theory and in turn in network analysis, since they essentially determine quite a number of mechanisms that can be developed over these underlying graphs-networks. For instance, in information networks, such vertices/edges control the information flow across the underlying graph components, and thus, they can be of key importance for health, military, political, and financial applications.

The following results provide some characteristic properties of such vertices/edges of a graph, as well as relations that can be used either for identifying these groups of vertices/edges, or for exploiting/taking them into account in the design and analysis of network mechanisms.

Theorem 9 *The minimum number of vertices separating two nonadjacent vertices s and t is the maximum number of disjoint s − t paths.*

The above theorem indicates the essential structure imposed in a graph by disjoint paths and also, through the following theorem, it provides insight in the relation of disjoint paths with n-components:

Theorem 10 *A graph is n-connected if and only if every pair of points are joined by at least n vertex-disjoint paths.*

Some additional and useful results on disjoint paths follow:

Theorem 11 *For any two vertices of a graph, the maximum number of edge-disjoint paths joining them equals the minimum number of edges that separate them.*

This theorem can be used not only in mathematical proofs regarding bounds, but more importantly in practical applications, where disjoint path discovery is required, e.g., information or metabolic pathways in data and biological networks, respectively, for obtaining useful estimations on disjoint connecting paths, especially in cases of large network populations.

An immediate outcome of the above theorem, which is intuitively expected, is that provided by the following theorem:

Theorem 12 *A graph is n-edge-connected if and only if every pair of vertices are joined by at least n edge-disjoint paths.*

which provides a way for determining if a network is n-edge-connected.

Furthermore, the above results can be extended in the following theorem to cover cases of subsets of vertices, rather than only individual nodes as in the previous results:

Theorem 13 *For any two disjoint nonempty sets of vertices V_1 and V_2, the maximum number of disjoint paths joining V_1 and V_2 is equal to the minimum number of vertices that separate V_1 and V_2.*

Alternatively, considering subsets of vertices and their connecting paths, one may obtain results regarding the connectivity of a graph. The following theorem is a characteristic example:

Theorem 14 *A graph with at least 2n vertices is n-connected if and only if for any two disjoint sets V_1 and V_2 of n vertices each, there exist n disjoint paths joining these two sets of vertices.*

Theorem 15 *The ordered pair (a, b) is a connectivity pair for vertices u and v in a graph G if and only if there exist a vertex-disjoint $u - v$ path and also b edge-disjoint $u - v$ paths, which are the maximum possible numbers of such paths.*

A closely related concept in Graph Theory, which is relevant not only to connectivity, but also to notions such as paths and flow is that of cuts and cut-sets.

Definition 11 *A cut is a partition of the vertices of a graph into two disjoint subsets.*

The *cut-set* of the cut is the set of edges whose end points are in different subsets of the partition, as shown in Figure 2.7. Edges are said to be crossing the cut if they are in its cut-set. In an unweighted undirected graph, the size or weight of a cut is the number of edges crossing the cut. In a weighted graph, the same term is defined by the sum of the weights of the edges crossing the cut.

Theorem 16 *In any graph, the maximum number of edge-disjoint cutsets of edges separating two vertices u and v is equal to the minimum number of edges in a path joining u and v.*

The notion of path will be described in more detail in the following section. Intuitively, a path consists of a sequence of edges and vertices in a graph.

Apart from the results regarding connectivity, results that characterize loss of connectivity are also important in network science, especially those describing the loss of connectivity in the limit, namely when a minor modification in

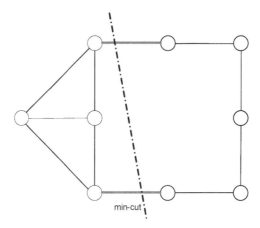

Figure 2.7: Example of cut and a cut-set.

the network leads the graph from connectivity to disconnected. The latter has been extensively studied in the context of percolation theory [34], [77], [150] where other phase transition phenomena have also been studied, apart from connectivity. Within the previously described framework of information flow, network analysis will be required to identify the critical resources, and graph components, loss of which can potentially harm the operations of the networks. The following concepts are relevant to these concerns in network analysis.

A *cutpoint* (cutvertex) of a graph is one whose removal increases the number of components. A *bridge* is such an edge increasing the number of components of the graph.

A *nonseparable* graph is connected, non-trivial, and has no cutpoints. A *block* of a graph is a maximal nonseparable subgraph. If graph G is nonseparable, then G itself is often called a block.

The following theorems provides a straightforward characterization of graph potentials regarding cutpoints, bridges, blocks and cycles.

Theorem 17 *Let v be a vertex of a connected graph G. The following are all equivalent:*

1. *Vertex v is a cutpoint of G.*

2. *There exist vertices u and w distinct from v such that v is on every $u - w$ path.*

3. *There exists a partition of the set of vertices $V - \{v\}$ into subsets U and W such that for any vertices $u \in U$ and $w \in W$, vertex v is on every $u - w$ path.*

Theorem 18 *Let x be an edge of a connected graph G. The following are all equivalent:*

1. *Edge x is a bridge of G.*

2. *Edge x is not on any cycle of G.*

3. *There exist vertices u and v of G such that the edge x is on every path joining u and v.*

4. *There exists a partition of V into subsets U and W such that for any vertices $u \in U$ and $w \in W$, edge x is on every path joining u and w.*

Theorem 19 *Let G be a connected graph G with at least three vertices. The following are all equivalent:*

1. *G is a block.*

2. *Every two vertices of G lie on a common cycle.*

3. *Every vertex and edge of G lie on a common cycle.*

4. *Every two edges of G lie on a common cycle.*

5. *Given two vertices and one edge of G, there is a path joining the vertices that contain the edge.*

6. *For every three distinct vertices of G, there is a path joining any two of them, which contains the third.*

7. *For every three distinct vertices of G, there is a path joining any two of them, which does not contain the third.*

These results have been used extensively in various applications and implementation in Network Science and complex/social network analysis, especially in cases of grouping, structure discovery, and other occasions where the discovery of structure in a network is important for social-, robustness-, and network-related studies.

A final result ensures that for every connected graph that is nontrivial, not all nodes are cutpoints, thus the graph does not have trivial connections.

Theorem 20 *Every nontrivially connected graph has at least two points that are not cutpoints.*

2.1.4   Paths and Cycles

One of the important aspects in network science and analysis is the traversality potentials of a graph, which describe cumulatively all relevant properties and relations regarding the possibility of sequentially visiting nodes and edges of a graph within different application perspectives, such as communication (implicit or explicit), or flow (which will be analyzed in more detail in the following subsection) of information.

As explained, the "ability to traverse" describes the potential of sequential visiting of vertices and/or edges of the network. Thus, it is strongly connected to the connectivity (previous subsection), flow (following subsection), and covering/planarity aspects of graphs. From this, it is evident that path and cycles traversing/spanning a network can be tightly connected to the fundamental structural properties of the network, and thus they can be very indicative of the potentials of the specific network with respect to the application suitability, etc. The most fundamental concept regarding the notions of paths and cycles in graphs is that of a walk.

Definition 12 *A* walk *of a graph G is an alternating sequence of vertices and edges, e.g., $v_0, x_1, v_1, ..., v_{n-1}, x_n, v_n$, beginning and ending with vertices in which each edge is incident with the two vertices immediately preceding and following it in the given sequence.*

Such a walk essentially joins v_0 and v_n and could be simply denoted by $v_0, v_1, ..., v_{n-1}, v_n$ and referred to as a $v_0 - v_n$ walk. A walk is closed if $v_0 = v_n$, otherwise it is open. A walk is called a *trail* if all the edges of the walk are distinct. If in addition all the vertices (and thus necessarily all the edges) are distinct the walk is called a *path*. Another important notion is that of a *cycle*.

Definition 13 *A cycle is a closed walk with distinct vertices and it is denoted as C_n (cycle on n points, $n \geqslant 3$).*

Graph C_3 is called a triangle. A path with n points is denoted as P_n.

The length of a walk $v_0, v_1, ..., v_{n-1}, v_n$ is n, equal to the number of edges of the network traversed. The *girth* of a graph G, denoted by $g(G)$ is the length of the shortest cycle (if any) in G, while the *circumference* $c(G)$ is the length of the longest cycle (if any, as well).

The *distance* $d(u, v)$ between two vertices u, v in G is defined as the length of the shortest path joining them, if existing. Otherwise, the distance, defined as above, is considered $d(u, v) = \infty$. If the underlying graph G is connected, the distance is a proper metric, which means that for all points u, v, w the following properties are satisfied:

1. $d(u, v) \geqslant 0$, with $d(u, v) = 0$ if and only if $u = v$.

2. $d(u, v) = d(v, u)$

3. $d(u, v) + d(v, w) \geqslant d(u, w)$ (triangle inequality).

Usually, in measure theory, the shortest path between two nodes u-v is called *geodesic*. Thus, the *diameter* $d(G)$ of a connected graph G is the length of the longest geodesic. The diameter is indicative of the longest path one would experience in the best case scenarios where distances are computed according to geodesics. For instance, in routing in communications networks, the shortest paths are taken into account, and thus the diameter would be the maximum distance experienced in multihop packet routing.

If a graph G has a walk that traverses each edge exactly once and goes through all vertices, then G has an *Eulerian trail*. If the Eulerian trail is closed, then G is said to have an *Eulerian circuit* and G is called Eulerian. In other words, G is Eulerian, if it has an euler circuit. Clearly, a graph G cannot be Eulerian if it is disconnected. The following theorems reflect exactly this fact and offer further insight on the properties of Eulerian graphs:

Theorem 21 *The following statements are equivalent for a connected graph G:*

1. *G is Eulerian.*

2. *Every vertex of G has even degree.*

3. *The set of edges of G can be partitioned into cycles.*

Similarly to Euler trails and circuits, a *Hamilton cycle* is a cycle containing all the vertices of a graph. A *Hamilton path* is a path containing all the vertices of a graph. A *Hamilton graph* is a graph containing a Hamilton cycle. Thus, a Hamilton graph has a closed Hamilton path.

The notion of Hamilton cycles and especially paths is important for various applications, especially operations research. Unfortunately, no significant characterization exists for Hamilton graphs and in fact we still lack an efficient algorithm for constructing a Hamilton cycle in a Hamiltonian graph. Several necessary or sufficient conditions are available for the characterizations, such as the following theorem:

Theorem 22 *Every Hamiltonian graph is 2-connected.*

However, it is not yet known whether it is impossible to eventually obtain efficient algorithms for constructing Hamilton cycles in graphs. Such a result would be rather prominent and useful in practical applications of operations research and more.

2.1.5   Flow

Another important aspect of the macroscopic operation of a network, second only to the most fundamental of connectivity, is the transfer of some type of quantity from a member of the network to other members through the allowed paths (those imposed by connectivity). The transferred quantity could be of various types, depending on the application framework. For instance, in communications networks it could be data, in social networks it could be news or another type of information, in financial networks it could be cash flow, etc. Intuitively, such a notion is similar to water flow in a pipe network and a mapping of the water intensity to the flow of information, currency, etc., is possible to aid in the analysis and study of such problems.

In this subsection, we focus on flow-related aspects of networks, which essentially describe the transferring capabilities of networks and the related properties. We start with necessary definitions and then proceed with properties and features developed. In this case, since the direction of flow is important (according to the water flow analogy the direction of flow is critical for the water transferring capabilities of the pipe network), the underlying graphs assumed for the study of flow need to be considered as directed. Such graph types are considered in the sequel, unless otherwise noted.

Assume a finite directed graph $\overrightarrow{G}(V, \overrightarrow{E})$ with two special vertices, namely the *source* s and the *sink* t, where $\overrightarrow{E}$ is a directed set of edges (oftentimes referred to as arcs).

Definition 14 *A flow is defined as a nonnegative function on the edges, where the value $f(\overrightarrow{xy}) = f(x, y)$ is the amount of flow traversing the edge-arc $\overrightarrow{xy}$.*

We also assume that $f(x, y) = 0$ whenever $\overrightarrow{xy} \notin \overrightarrow{E}$.

One of the basic principles in nature is that the total flow into each intermediate vertex equals the total flow leaving the vertex (Kirchhoff's law [33]), except for the source and sink nodes. Using the water flow analogy, this essentially may be thought of as the fact that in pipe junctions, the amount of water towards the junction has to equal the amount of water out of the junction, provided the junction is no source/sink. If that was not the case, the amount of water remaining in the junction would either increase rapidly and eventually blow up, or some amount of water would have to be created out of nothing. This is essentially an application of the law of conservation in nature. It may be expressed as:

$$\sum_{y \in \Gamma^+(x)} f(x,y) = \sum_{z \in \Gamma^-(x)} f(z,x) \qquad (2.3)$$

for all $x \in V - \{s,t\}$ and the outgoing flow neighborhood of x, $\Gamma^+(x) = \{y \in V : \overrightarrow{xy} \in \overrightarrow{E}\}$, incoming flow neighborhood $\Gamma^-(x) = \{y \in V : \overrightarrow{yx} \in \overrightarrow{E}\}$. At the same time, the total flow leaving the source equals the total flow entering the sink:

$$\sum_{y \in \Gamma^+(s)} f(s,y) - \sum_{y \in \Gamma^-(s)} f(y,s) = \sum_{y \in \Gamma^-(t)} f(y,t) - \sum_{y \in \Gamma^+(t)} f(t,y) \qquad (2.4)$$

corresponding to a network, where no external input is assumed and the only flow is generated from the source. Exactly this flow has to eventually reach the sink, since the intermediate network nodes do not generate, nor absorb any flow. The net flow value from s to t is denoted by $v(f)$.

With each edge $\overrightarrow{xy} \in \overrightarrow{E}$, we associate a nonnegative number $c(x,y)$, called the *capacity* of $\overrightarrow{xy}$, corresponding to the description that the flow of this edge cannot exceed the capacity for any reason. Given two subsets X, Y of V, $\overrightarrow{E}(X,Y)$ is the set of directed $X - Y$ edges $\overrightarrow{E}(X,Y) = \{\overrightarrow{xy} \in \overrightarrow{E} : x \in X, y \in Y\}$. Whenever $g : \overrightarrow{E} \to \mathbb{R}$ we put $g(X,Y) = \sum g(x,y)$, where the summation is over $\overrightarrow{E}(X,Y)$. If S is a subset of V containing s but not t, then $\overrightarrow{E}(S,\bar{S})$ is called a cut separating s from t. The capacity of the cut $\overrightarrow{E}(S,\bar{S})$ is defined as $c(S,\bar{S}) = \sum c(x,y)$, with the summation spanning the set $\overrightarrow{E}(S,\bar{S})$.

By definition, edge capacity and flow cut are rather crudely related to $v(f) \le \sum_{\overrightarrow{xy} \in \overrightarrow{E}} c(x,y)$.

Perhaps the most important result regarding network flow, which is widely employed and exploited in many problem solving approaches, is the Max-Flow Min-Cut theorem. The Max-Flow Min-Cut theorem allows identifying the edges of a network that determine the flow dynamics in the network. It can be expressed as:

Theorem 23 *(Max-Flow Min-Cut Theorem) The maximal flow value from s to t is equal to the minimum of the capacities of cuts separating s from t.*

The theorem remains valid even if some edges have infinite capacity, but the maximal flow value is finite.

On one hand, the Max-Flow Min-Cut theorem enables the computation of the maximum transferring capacity of the network, while on the other, it enables us to locate the bottleneck part of the graph restricting this capacity. Both are important for the analysis of operation and the design of efficient networks.

An important fact regarding the structural properties of the capacity functional is provided by the following theorem:

Theorem 24 *If the capacity function is integral then there is a maximal flow that is also integral.*

which essentially describes the behavior of the maximal flow given the behavior of the capacity function.

For the cases with multiple source and (or) sink nodes, the max-flow min-cut theorem is straightforwardly extended as:

Theorem 25 *The maximum of the flow value from a set of sources to a set of sinks is equal to the minimum of capacities of cuts separating the sources from the sinks.*

The notion of capacity can be extended to cover vertices as well. More specifically, assume capacity restrictions defined over all vertices except the source and sink nodes. Thus, a vertex capacity constraint will be a function $c : V - \{s, t\} \to \mathbb{R}^+$ and every flow f from s to t has to satisfy the following inequality for all $x \in V - \{s, t\}$:

$$\sum_{y \in \Gamma^+(x)} f(x, y) = \sum_{z \in \Gamma^-(x)} f(z, x) \le c(x) \qquad (2.5)$$

In this case, a cut is a subset S of $V - \{s, t\}$, such that no positive-valued flow from s to t can be defined on $G - S$. The max-flow mix-cut theorem can also be cast in a vertex form as follows:

Theorem 26 *Let $\vec{G}$ be a directed graph with capacity bounds on the vertices other than the source s and sink t. Then the minimum of the capacity of a vertex-cut is equal to the maximum of the flow value from s to t.*

2.1.6   Planarity

Another facet of graphs, which is very useful especially in the application domain, such as topography, location-based applications (geo-location), and visualization applications, is the topology of the graph in terms of embedding it in specific spaces, e.g., a plane or a sphere.

A graph is said to be embedded in a surface S when it is drawn on S so that no two edges intersect.

Definition 15 *A graph is* planar *if it can be embedded in the plane.*

A plane graph has already been embedded in the plane. Regions defined by a plane graph are referred to as *faces* and the unbounded region is called the *exterior face*.

Planar graphs are very important since numerous network behaviors are developing in surfaces, e.g., cell membranes in biology, earth in communications, and social networks, materials, etc. Apart from the evident applications, and some others not so evidently expected, planarity offers significant insight on the characteristics and capabilities of a network, in addition to the rest of its analyzed properties. Several graph properties are tightly related to planarity and various results on other features of a graph can be also derived through results involving planarity.

Perhaps the most important result regarding planarity and graphs is the following:

Theorem 27 *(Euler Polyhedron Formula) For any spherical polyhedron with p vertices, q edges, and F faces, $p - q + F = 2$.*

A *plane map* is a connected plane graph together with all its faces. A relevant result connecting plane graphs with graph cycles, which can be used in various applications for identifying plane maps (e.g., topography and roadmap networks) is the following:

Corollary 2 *If G is a (p, q) plane map in which every face is an n-cycle, then $q = n(p - 2)/(n - 2)$.*

A maximal planar graph is one in which no line can be added without losing planarity. The following corollary provides a characterization of maximal planar graphs regarding the types of their faces.

Corollary 3 *If G is a (p, q) maximal plane graph, then every face is a triangle and $q = 3p - 6$. If G is a plane graph in which every face is a 4-cycle, then $q = 2p - 4$.*

Some useful and relatively simple results that can be exploited in various applications are the following. They essentially provide tools for identifying graphs that can be planar or non-planar and can be used for designing desirable mechanisms for various types of networks and application frameworks.

Corollary 4 *If G is any planar (p, q) graph with $p \geq 3$, then $q \leq 3p - 6$. Furthermore, if G has no triangles, then $q \leq 2p - 4$.*

Corollary 5 *The graphs K_5 and $K_{3,3}$ are nonplanar.*

Corollary 6 *Every planar graph G with $p \geq 4$ has at least four vertices of degree not exceeding 5.*

Perhaps one of the most important and more useful tools for identifying planar networks is the following theorem. It can be very easily—and sometimes efficiently as well—applied through greedy algorithms and used in practical applications.

Theorem 28 *A graph is planar if and only if each of its blocks is planar.*

This means that planarity is not only a global graph property, but rather it characterizes the graph in smaller scales as well. A graph is planar as long as the smaller parts of it are also planar and vice-versa. This 'recursive'-like feature of planarity can be exploited in many applications related to planarity, regarding both ways, namely top-down or bottom-up.

The following set of results provides useful tools for various applications related to networks and planarity, e.g., roadmap networks, protein folding, etc.

Theorem 29 *Every 2-connected plane graph can be embedded in the plane so that any specific face is the exterior.*

Corollary 7 *Every planar graph can be embedded in the plane so that a prescribed edge is an edge of the exterior region.*

Theorem 30 *Every maximal planar graph with $p \geq 4$ vertices is 3-connected.*

Theorem 31 *Every 3-connected planar graph is uniquely embeddable on the sphere.*

The following theorem is essentially considered the pinnacle of planarity, since it describes very-easy-to-understand conditions for characterizing a graph as non-planar.

Theorem 32 *(Kuratowski's theorem) A graph is planar if and only if it has no subgraph homeomorphic to K_5 or $K_{3,3}$.*

Thus, given Kuratowski's theorem, the following result can be derived.

Theorem 33 *A graph is planar if and only if it does not have a subgraph contractible to K_5 or $K_{3,3}$.*

The above result is very useful, since the problem of graph planarity characterization is reduced to one of locating K_5 or $K_{3,3}$ components in a graph. Greedy approaches may be developed for this purpose. However, depending on the specific structure of a graph, such special features maybe exploited for quicker and more efficient planarity characterization.

2.1.7 Coloring (Covering)

Another important facet of traditional Graph Theory is covering, sometimes equivalently denoted by coloring. In several emerging problems of traditional and modern complex networks, it is required to distinguish the vertices of a graph (i.e., complex network) in disjoint sets, in which nodes of the same set are non-neighboring. These disjoint sets may be considered to represent different color classes, so that the color corresponding to each class is used to color the vertices of the class and eventually no two vertices of the same color are adjacent. The latter is denoted by the vertex coloring problem, and due to duality, the problem can be similarly considered for network edges, in which case the problem is denoted by line (edge) coloring and no two adjacent lines should be colored by the same color. The alternative terminology employed refers to the objective of covering graph nodes with different colors and it is equivalently used.

In the vertex coloring case, the different classes of colors essentially define *independent sets* of nodes of the network. Such sets have multiple applications in complex networks, e.g., scheduling in wireless ad hoc networks, or voter groups of political parties in social networks. Examples are depicted in Figure 2.8, where vertex colors have been denoted by colored circles and edge colors have been denoted by directly coloring the corresponding edges with different colors. A maximum independent set is a largest independent set for a given graph G and its size is denoted by $\alpha(G)$ and referred to as *independence number* of G.

Both covering and coloring problems emerge very often, sometimes implicitly, and constitute critical factors for many emerging behaviors in various applications frameworks. In this subsection, we will present the fundamental definitions providing a solid basis for understanding some useful results characterizing coverings and graphs, which could be useful for the interested researcher and involved student.

Coloring refers to the assignments of different colors to the vertices (edges) of a network in such a way that adjacent vertices (edges) have distinct colors (Figure 2.8). The minimal number of colors in a vertex coloring is referred to as *chromatic number* $\chi(G)$, and the minimal number of colors in an edge coloring is referred to as *edge-chromatic number* $\chi'(G)$. It should be noted that $\chi'(G)$ is exactly the chromatic number of the line (edge) graph of G, $\chi'(G) = \chi(L(G))$. An *n-coloring* of a graph G uses n colors and thus, it partitions the edge set V into n color classes. Clearly, a graph G s *n-colorable* if $\chi(G) \leq n$ and it is *n-chromatic* if $\chi(G) = n$.

A straightforward consequence of the definition of the chromatic number is that it is at least as large as the maximum clique size of the network, i.e., $\chi(G) \geq \omega(G)$. Similarly, the edge-chromatic number is at least as large as the maximal network degree $\chi'(G) \geq \Delta(G)$.

Thus,

Theorem 34 *Let G be a connected graph with maximal degree Δ. Suppose G*

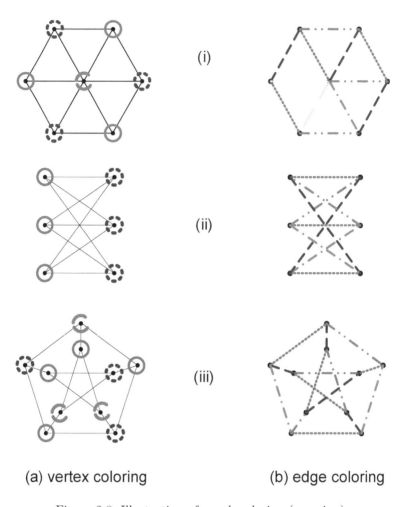

(i)

(ii)

(iii)

(a) vertex coloring   (b) edge coloring

Figure 2.8: Illustration of graph coloring (covering).

is neither a complete graph nor an odd cycle. Then $\chi(G) \leq \Delta$,

which leads to the following theorem:

Theorem 35 *A graph G of maximal degree* Δ *has edge-chromatic number* Δ *or* $\Delta + 1$.

A more involved result is described in the following theorem:

Theorem 36 *Every plane graph is 5-colorable.*

which relates colorability with planarity and can be useful in various practical network applications, ranging from topography and roadmap networks to protein folding networks and socio-politics dynamics.

A more advanced notion in colorability is that of perfect graphs. A graph G is *perfect* if $\chi(H) = \omega(H)$ for every induced subgraph H of G, including G itself.

Perfect graphs are closely related to bipartite graphs as the following theorem asserts.

Theorem 37 *The complement of a bipartite graph is perfect.*

Thus,

Theorem 38 *A graph is bicolorable if and only if it has no odd cycles.*

which is derived straightforwardly from the properties of bipartite graphs.

The complements of perfect graphs are also perfect graphs as asserted by the following theorem.

Theorem 39 *The complement of a perfect graph is perfect.*

This means that a bipartite graph is also perfect. In fact, even more can be obtained, so that every bipartite graph, the complement of every bipartite graph, the line graph of every bipartite graph, and the complement of the line graph of every bipartite graph, are all perfect.

A graph G is perfect if and only if neither G, not its complement $\bar{G}$ contains an induced odd cycle of length at least 5.

One of the most important results in Graph Theory is the four color conjecture, which was eventually proven and turned into the following theorem:

Theorem 40 *(Four Color Theorem) Every planar graph is 4-colorable.*

We note that graph coloring here is meant in the sense of vertex coloring. The Four Color Theorem relates planarity and colorability in a simple fashion. It also limits the colorability potentials for a graph, namely 4 is the minimum number of colors one will require to color the vertices of a planar graph, in the general case. Thus, even though less than 4 colors might be sufficient, e.g., as defined by the next theorem:

Theorem 41 *Every planar graph with fewer than 4 triangles is 3-colorable.*

there are special cases, as is evident by the above restriction of 4-triangle block absence. Several other similar results are available, which are not presented here due to their specialized nature.

2.1.8 Algebraic Graph Theory

A graph can be completely defined by its vertex or edge incidences and luckily, both can be efficiently represented and exploited in matrix forms. Algebraic Graph Theory, which among others, studies the properties of these representation matrices, is a field of its own merit and its in-depth description is out of the scope of the field of complex networks (however, it is a mathematical

The Lagrangian $f(G)$ is intimately related to the complete subgraphs of G. More specifically, it depends only on the clique number of G. If G is a complete graph K_n, then $f(K_n) = \frac{n-1}{n}$.

Theorem 44 *Let G be a graph with clique number k_0. Then $f(G) = \frac{k_0-1}{k_0}$.*

The combinatorial Laplacian of G is simply $L = D - A$, where $D = [D_{ij}]$ is the diagonal matrix in which D_{ii} is the degree $d(v_i)$ of node v_i. The Laplacian is a matrix as well. The second smallest eigenvalue of L is especially important. The larger this eigenvalue is, the tougher it is to cut G into pieces and the more G expands (see Expander Graphs at Chapter 5, Section 5.5). The Laplacian is an even more powerful tool than the Lagrangian. If G is r-regular, the spectrum of L is just the spectrum of G reversed and shifted.

The Laplacian is useful in characterizing various connectivity features of the graph with other measures, as the following results show:

Theorem 45 *The vertex connectivity of an incomplete graph G is at least as large as the second smallest eigenvalue $\lambda_2(G)$ of the Laplacian G.*

Theorem 46 *The adjacency matrix of a graph G has at least $\alpha(G)$ non-negative and at least $\alpha(G)$ non-positive eigenvalues, counted with multiplicity, where $\alpha(G)$ is the independence number of graph G.*

Obtaining knowledge of the entire eigenvalue spectrum of a graph would be ideal for the design, study, and analysis of the systems where the graphs emerge. However, this is intractable, if not impossible, in most applications. Various bounds on the eigenvalue spectrum constitute a possible alternative, whereas in some special cases, special graph topological properties could be exploited to obtain sufficient or complete spectrum knowledge.

Trivially, the empty graph $E_n = \bar{K}_n$ has one eigenvalue 0, with multiplicity n, and more generally, adding an isolated vertex to a graph G increases by one the multiplicity of the 0 eigenvalue. The complete graph K_n has eigenvalues $\mu_1 = n - 1$ and $\mu_2 = -1$, with multiplicities 1 and $n - 1$ respectively. The complete bipartite graph $K_{n,n-k}$ has three eigenvalues $\mu_1 = (k(n-k))^2$ and $\mu_2 = -(k(n-k))^2$, each with multiplicity 1 and $\mu_3 = 0$ with multiplicity $n-2$.

At this point, we quote the Perron–Frobenius Theorem [130], which characterizes the eigenvalues and eigenvectors of nonnegative, irreducible matrices and finds interesting applications in complex networks. A matrix A is said to be nonnegative if it consists of nonnegative entries, i.e., $A = [a_{ij}]$, where $a_{ij} \leq 0$, $\forall\, i, j$. A matrix A is said to be reducible if it is possible to rewrite A after row and column operations so that its upper right-hand block consists of zero elements, i.e.,

$$A = \begin{pmatrix} B & 0 \\ C & D \end{pmatrix},$$

where B, D are square matrices. The Perron–Frobenius Theorem refers to irreducible matrices, thus matrices not being reducible. Also, the spectral radius of A is defined as $\rho(A) = \max_i\{|\mu_i|\}$, i.e., $\rho(a)$ coincides with the eigenvalue of A with the largest absolute value.

Theorem 47 *(The Perron–Frobenius Theorem) If A is a nonnegative irreducible matrix, then:*

1. *$\rho(A) > 0$ and $\rho(A)$ is an eigenvalue of A, i.e., the largest eigenvalue of A is always positive and it equals the spectral radius of A.*

2. *The left and right eigenvectors of A corresponding to the eigenvalue $\rho(A)$ consist of positive entries (i.e., positive vectors).*

3. *If μ is any other eigenvalue of A, then $|\mu| \leq \rho(A)$. Equality of this relation is achieved in the case that A is a periodic matrix with period T, i.e., $a_{ii}^n = 0$, $n \neq kT$.*

The Perron–Frobenius Theorem has plenty of applications in the field of complex networks [130]. It is used in the power control problem where multiple transmitting nodes need to identify their optimal transmission powers so as to achieve successful communication under acceptable interference levels, in commodity pricing, in population growth models, in the page ranking algorithm used by the Google search engine, and in the expander graph structure.

2.2 Random Graphs

Random Graphs are considered one of the most contemporary branches in Graph Theory, essentially started by a series of papers by Erdős and Rényi in the 1940s and 1950s. Using probabilistic methods several unexpected results on graphs were proven, some of which are deterministic. It was possible to demonstrate the existence of the desired graphs without actually constructing them, using probabilistic methods only.

Compared to traditional Graph Theory, the most important results of which were summarized in the previous parts of this chapter, random graphs study essentially the same problems from another perspective, more suitable for studying the dynamic behavior of networks. Random graphs also offer an alternative network model for cases where stochastic behavior is prevalent. In the latter, various degrees of randomness can be incorporated in the formation and evolution of design of networks, which was not straightforwardly achieved in the case of traditional Graph Theory models.

An intuitive way of considering Random Graphs is that every probability space whose points are graphs provides such a notion of a random graph. However, this also gives rise to many different Random Graphs models. In the following, we will summarize the most important ones and provide the more useful results stemming from them.

2.2.1   Basic Random Graph Models

In general, one encounters three basic and closely related models of Random Graphs in the literature. The probability space in each case consists of graphs on a fixed set of n distinguishable vertices $V = [n] = \{1, 2, ..., n\}$. In the sequel, we use the notations M, N to represent number of edges.

For $0 \leq M \leq N$, the space $\mathcal{G}(n, M)$ consists of all $\binom{N}{M}$ subgraphs of K_n with M edges. Thus, the $\mathcal{G}(n, M)$ model describes graphs that could be obtained by the subspace of K_n which contains only graphs with M edges. All elements (graphs) of this space are assumed equiprobable, and due to the assignment of a probability measure to its elements, the space becomes a probability space. It is customary to write $G_M = G_{n,M}$ for a random graph in the space $\mathcal{G}(n, M)$. The probability that G_M is precisely a fixed graph H on $[n]$ with M edges is:

$$\mathbb{P}_M(G_M = H) = \binom{N}{M}^{-1} \tag{2.8}$$

The space $\mathcal{G}(n, p)$ ($\mathcal{G}(n, \mathbb{P}(edge) = p)$) is defined for probability $0 \leq p \leq 1$. A random element of this space corresponds to selecting edges independently with probability p, for all possible existing edges in a graph of n nodes. This means again that the potential probability space includes all possible subgraphs of the space of K_n graphs. In this case, however, only those are selected that correspond to a selection process where each edge is selected with probability p. Similarly to the $\mathcal{G}(n, M)$ model, the probability of a fixed graph H on $[n]$ with m edges is:

$$p^m (1 - p)^{N-m} \tag{2.9}$$

where each of the m edges of H has to be selected, and none of the $N - m$ is allowed to be selected.

In most cases, in Random Graphs one is interested in the behavior of graphs for $n \longrightarrow \infty$. Thus, both $M = M(n)$ and $p = p(n)$ for the two models become functions of n and so do most of the studied properties of the graphs.

Finally, the space $\mathcal{G}(n, 1/2)$ is obtained by picking one of the 2^N graphs on $[n]$ at random, thus it is the space of Random Graphs of order n, where all graphs are considered equiprobable.

The three spaces are very closely related to each other. For $M \sim pN$ the spaces $\mathcal{G}(n, M)$ and $\mathcal{G}(n, p)$ are close to each other.

Once a probability space on Random Graphs has been defined, every graph invariant, such as the ones defined for traditional deterministic graphs, becomes a random variable. These random variables depend on the specific random space employed.

It has been proven that $\mathcal{G}(n, M)$ and $\mathcal{G}(n, p)$ are practically interchangeable in many cases, provided that $M = pN$, as $M \to \infty$ and $(N - M) \to \infty$.

Simple Properties of Almost All Graphs

For the rest of this subsection, unless otherwise noted explicitly, all reference will be with respect to the probability space $\Omega_n = \mathcal{G}(n, p)$ of random graphs. It could also be assumed that for the following results, $0 < p < 1$ is fixed, namely p is independent of n.

Given a property Q, it is customarily denoted that *almost every (a.e.)* graph in the probability space Ω_n consisting of graphs of order n has property Q if $\mathbb{P}(G \in \Omega_n : G \text{ has } Q) \to 1$ as $n \to \infty$.

One of the most simple properties of almost every graph is that if H is an arbitrary fixed graph, then almost every $G_P \in \mathcal{G}(n, p)$ contains H as a spanned subgraph. A stronger result of this is the following:

Theorem 48 *Let $1 \leq h \leq k$ be fixed natural numbers and let $0 < p < 1$ be fixed as well. Then in $\mathcal{G}(n, p)$ a.e. graph G_p is such that for every sequence of k vertices $x_1, x_2, ..., x_k$ there exists a vertex x such that $xx_i \in E(G_p)$ if $1 \leq i \leq h$ and $xx_i \notin E(G_p)$ if $h < i \leq k$.*

As noted before, $\mathcal{G}(n, p)$ and $\mathcal{G}(n, M)$ are practically interchangeable in many situations, provided that $p = M/N$, for $M \to \infty$ and $(N - M) \to \infty$. This can be expressed more strongly in the following theorem:

Theorem 49 *Let $0 < p = p(n) < 1$ be such that $pn^2 \to \infty$ and $(1-p)n^2 \to \infty$ as $n \to \infty$ and let Q be a property of graphs.*

- *Suppose $\epsilon > 0$ is fixed and if $(1 - \epsilon)pN < M < (1 + \epsilon)pN$, then a.e. graph in $\mathcal{G}(n, M)$ has Q. Then a.e graph in $\mathcal{G}(n, p)$ has Q.*

- *If Q is a convex property and a.e. graph in $\mathcal{G}(n, P(edge) = p)$ has Q, then a.e. graph in $\mathcal{G}(n, \lfloor pn \rfloor)$ has Q.*

The following is a characterization of the clique number through random graph tools:

Theorem 50 *Let $0 < p < 1$ and b fixed. Then the clique number of almost every $G_{n,p}$ is d or $d + 1$, where $d = d(n)$ is given by:*

$$\binom{n}{d} p^{\binom{d}{2}} \geq \log n \tag{2.10}$$

A property of a graph is essentially a class of graphs closed under isomorphism. In particular for random graphs, a property Q_n of graphs of order n can be viewed as a subset of the set of graphs with vertex set $[n]$. This set needs only be invariant under permutations of $[n]$. A property Q of graphs is thus *monotone increasing* if Q is invariant under the addition of edges. Similarly a property is *monotone decreasing* if it is invariant under the deletion of edges.

A very interesting finding in Random Graphs was that overall a monotone increasing property of graphs arises rather abruptly, which was coined as threshold behavior. To express this more precisely, a function $p_\ell(n)$ is a *lower threshold function (ltf)* for a monotone increasing property Q if almost no $G_{n,p_\ell}(n)$ has Q and $p_u(n)$ is an *upper threshold function (utf)* for Q if almost every $G_{n,p_u(n)}$ has Q. Threshold functions are defined similarly for the space $\mathcal{G}(N, M)$.

Threshold behavior in Random Graphs is closely related to connectivity. The first result is the following:

Theorem 51 *Let $\omega(n) \to \infty$ and set $p_\ell = (\log n - \omega(n))/n$ and $p_u = (\log n + \omega(n))/n$. Then a.e. G_{p_ℓ} is disconnected and a.e. G_{p_u} is connected. Thus, for the model $\mathcal{G}(n, p)$, p_ℓ is a utf and p_u is a utf for the property of being connected.*

Another important result is the emergence of connectivity in random graphs. Given a monotone increasing property Q, the time τ at which Q appears in a graph process $\tilde{G} = (G_t)_0^N$ is the *hitting time* of Q, defined as:

$$\tau = \tau_Q = \tau(\tilde{G}; Q) = \min\{t : G_t \text{ has } Q\}. \tag{2.11}$$

Relevant to the above is the very important theorem:

Theorem 52 *For almost every graph process $\tilde{G}$ we have $\tau(\tilde{G}; conn) = \tau(\tilde{G}; \delta \geq 1)$,*

where δ is the minimum node degree and *conn* denotes the connectivity property. This Theorem essentially means that if one starts with an empty graph on a large set of vertices and adds edges randomly until the graph has no isolated vertices, then with high probability, the graph obtained is connected. This also means that the edge that removes the last isolated vertex makes the graph connected. Such a finding is very important, especially in wireless communications networks, having a lot of implications, and its great value is also that it can be used conversely when studying loss of connectivity. The theorem essentially says that both properties are identical and reversible.

By observing that the complement of a random graph $G_{n,p}$ is also a random graph $G_{n,q}$ with $q = 1 - p$, the distribution of the independence number $\alpha(G_{n,p})$ is precisely the distribution of the clique number $\omega(G_{n,p})$, and since by the previous theorem (Theorem 51) $\omega(G_{n,p}) = (\frac{1}{2} + o(1)) \log n / \log(1/q)$, then we have the following result:

Theorem 53 *Let $0 < p < 1$ be fixed. Then*

$$\chi(G_{n,p}) \geq \left(\frac{1}{2} + o(1)\right) \frac{\log n}{\log(1/q)} \tag{2.12}$$

for a.e. $G_{n,p}$, where $q = 1 - p$.

Regarding Hamilton cycles, some considerable results follow. which can characterize graphs and their 'Hamiltonian' properties:

Theorem 54 *Let $p = (c \log n)/n$ and consider the space $\mathcal{G}(n, p)$. If $c > 3$ and x and y are arbitrary vertices, then almost every graph contains a Hamilton path from x to y. If $c > 9$ then almost every graph is Hamiltonian connected, thus every pair of distinct vertices is joined by a Hamilton path.*

This is evidently a very useful result for information networks and the spread/propagation of information analysis in military, financial, and commercial perspectives. Similarly to connectivity threshold behavior, regarding Hamilton paths, we have:

Theorem 55 *Almost every graph process $\tilde{G}$ is such that $\tau(\tilde{G}; Ham) = \tau(\tilde{G}; \delta \geq 2)$, where "Ham" is the property of being Hamiltonian and "$\delta \geq 2$" is the property of having minimal degree at least 2.*

Consequently, if the random graph process is stopped as soon as the last vertex with degree at most 1 exists (it becomes a vertex with degree more than 1), then with high probability a Hamilton graph is obtained.

Finally, a very useful result regarding the components of a random graph and essentially providing an idea of how a typical graph looks is the following:

Theorem 56 *Almost every random graph process is such that if $k \geq 2$ is fixed and $t = o(n^{(k-1)/k})$ then every component of G_t is a tree of order at most t. Furthermore, if s is constant and $t/n^{(k-1)/k} \to \infty$ then G_t has at least s components of order k.*

2.3   Notation

The following Table 2.1, presents cumulatively the most important notation symbols regarding the theory of graphs and random graphs. The purpose of this table is to provide a solid nomenclature for the rest of this book and future studies of the interested reader and progressively develop a concrete idea of quantities of interest in the broader field of network graphs, which can be extended in social and complex network analysis.

Notation adopts the most frequently encountered paradigms, e.g., the order of a graph $G(V, E)$—namely the number of vertices of a graph is commonly denoted by $n = |G|$, and Table 2.1 reflects such common practice. On the other hand, the neighborhood of a vertex x is usually denoted by $\Gamma(x)$ for mathematicians and $N(x)$ by engineers. In any case, Table 2.1 includes both notations (but does not indicate which is used in each case).

Table 2.1: Graph theory notation.

Symbol	Definition
n	order (number of vertices) of $G(V, E)$ $(n = |V|)$
m	size (number of edges) of $G(V, E)$ $(m = |E|)$
$A = A(G)$	adjacency matrix of $G(V, E)$
$\alpha(G)$	independence number of $G(V, E)$
$B = B(G)$	incidence matrix of $G(V, E)$
$C(s)$	cover time
$C(x, y)$	mean commute time between vertices x-y
$\chi(G)$	chromatic (vertex) number of $G(V, E)$
$\chi'(G)$	edge chromatic number of $G(V, E)$
Δ	maximum node degree of $G(V, E)$
δ	minimum node degree of $G(V, E)$
$diam(G)$	diameter of of $G(V, E)$
$E(U, W)$	set of U-W edges
$E(G)$	set of edges of $G(V, E)$
$H(\omega)$	Hamiltonian
$H(x, x)$	mean return time
$H(x, y)$	mean hitting time
K_n	complete graph of order n
$K_{p,q}$	complete bipartite graph
$\kappa(G)$	connectivity of $G(V, E)$
$\lambda(G)$	edge-connectivity of $G(V, E)$
$L(G)$	line graph of $G(V, E)$
$N(x)$ or $\Gamma(x)$	neighborhood of vertex $x \in V$ of $G(V, E)$
$V(G)$	vertex set of of $G(V, E)$
$\omega(G)$	clique number of $G(V, E)$

Chapter 3

Cognitive Methods and Evolutionary Computing

Evolutionary computing and various cognitive methods included within the broader framework of evolutionary approaches constitute a vast branch of computer science, where inspiration has been drawn from the processes of natural evolution, in a fashion similar to that in which communications networks can employ approaches inspired by social networking and vice-versa. The fundamental metaphor of evolutionary computing and cognitive methods is a characteristic style of problem solving, which is essentially an advanced trial-and-error search. Such an approach usually takes place in a stochastic manner, which ensures rapid response time and efficient implementation, while simultaneously ensuring sufficient exploration of the corresponding search space of possible solutions to the problem.

In this chapter we initially explain the concept of evolutionary computing and the underlying "evolutionary cycle" based on which the corresponding approaches operate. In the sequel, we present in detail the most characteristic evolutionary approach, namely genetic algorithms, with which we demonstrate the specific operations and interactions of the evolutionary algorithm modules. Then, we present other prominent evolutionary algorithms in a similar, but more concise manner compared to the one employed for genetic algorithms.

The features presented in this chapter essentially constitute broader guidelines for developing evolutionary algorithms. They could be employed and implemented in different ways in designing evolutionary algorithms. In the following chapters, we show how to exploit concepts from the approaches presented in this chapter for developing evolutionary topology modification/control mechanisms.

3.1   Brief History of Evolutionary Computing

According to [59], evolutionary computing can be summarized as applying Darwinian principles to automated problem solving (computing), usually with the aid of computers nowadays, even though this concept can be traced back to the early 1940s, before the breakthrough of computers as we perceive them today. Alan Turing was among the first to propose the concept of genetical (evolutionary) search (1948) and Bremermann had already executed computer experiments on optimization through evolution and recombination[1] by 1968 [65].

The first widespread attempts appeared in the 1960s, where evolutionary programming emerged and genetic algorithms (by Holland [87]) emerged in the United States, while in Germany, Rechenberg [132] and Schwefel [144] invented evolution strategies. These approaches eventually converged and from the 1990s on they are considered as representatives of the same technology, namely evolutionary computing. Genetic programming, which emerged in the 1990s as well, complemented the other three branches, and all of them are currently part of the literature referred to as evolutionary computing/evolutionary algorithms.

Since the mid 1980s and later from the 1990s, more systematically, several conferences, workshops, and other relevant events have been devoted to the subject of evolutionary computing and the topics spanned by it. Later, from the mid 1990s onwards, the first journal devoted to the subject appeared and since then, several others followed. A significant number of publications are nowadays annually devoted to evolutionary computing and several approaches shaped and developed within this framework can be encountered in publications of other disciplines as well, signifying the importance and applicability of such approaches in engineering and other scientific disciplines.

3.2   Elements from Evolution Theory

Observing natural evolution of human or other populations, the spontaneous behavior of these groups is centered around survival (usually referred to as reproduction for living species). Survival is constrained by the environment and it is achieved by a measure of the *population fitness*, namely a measure of the successfulness of the population effort and developed capabilities to adapt to environmental changes and succeed in achieving their goals. In that sense, the fitness is also a measure of the population quality, determining which of these populations has greater odds to survive (and to what degree) drastic or slower changes in their environment.

Considering macroscopically the natural evolution, natural selection plays a key role. Given constrained population survival, where selection plays a fundamental role, natural selection seems to favor those individuals that com-

---

[1]Evolution and recombination are basic evolutionary computing modules, which will be analyzed in the rest of this chapter.

pete for resources in the most effective manner. This is sometimes referred to as "survival of the fittest." Such competition-based selection is one of the two cornerstones of evolutionary processes and will be a significant factor in evolutionary computing as well.

In the context of the stochastic trial-and-error approach of evolutionary computing described above and given the principle of the survival of the fittest, for each problem a number of available (candidate) solutions exist, each characterized by different quality (i.e. how well they solve the problem). Solution quality (similar to population fitness in natural evolution) determines the chance that these solutions will be still considered and used as seeds for further candidate solutions, until the final solution is achieved.

As a trial-and-error approach, evolutionary approaches can be customarily cast as optimization problems over search spaces with the objective function to maximize or minimize a function. In the case that fitness is employed as the optimization objective, maximization is the usual approach (unless fitness of a malign population is employed, in which case a minimization would be desired more). If each population's objective function value (fitness or other) is represented in a suitable space, local optima and global optima signify the evolution of the population (problem solutions) and its effectiveness. Such optima (local or global) denote populations (solutions) that are better than all their neighboring populations (solutions). A problem in which there is only one population (solution point) that is fitter than all of its neighbors is known as unimodal and in the event of multiple populations (solution points) with the same fitness value, the problem is known as multi-modal.

Even though the link between evolution and optimization from the above discussion seems to be straightforward, it can also be misleading, as the evolutionary processes are not always monotone (unidirectional uphill in a maximization case). Since the process is essentially of a stochastic trial-and-error nature, and the population has finite size, while some choices are made at random, there exist cases where highly fit individuals can be lost from one population generation to another (contrary to optimization where local optima with highest fitness always survive local optima with lower objective values), or the whole population may face a loss of great variety affecting it considerably. Such behavior is referred to as *genetic drift* in evolutionary terms. The combined effect of drift and selection enable populations to move up and down the fitness scale, which in optimization terms enables the population to escape from local optima. The objective of an evolutionary approach would be to exploit this behavior to avoid trapping in local optima, while guaranteeing convergence to global ones when available and when desired.

At this point we need to clarify two aspects of evolution theory, namely genotype and phenotype, which will aid in understanding the process and flow of evolutionary computing. In genetics, each entity consists of both genotype and phenotype. *Genotype* contains all the information necessary for building an entity, while *phenotype* describes the outside (visible) properties of an entity. Thus, the genotype encodes the phenotype of an individual entity. The *genome* describes cumulatively the whole genetic information, namely the complete building plan of an entity. One of the principal dogmas of genetics is that information flow is one-way, namely that genotypic information only influences phenotypic and not vice-versa. Consequently, the genetic material of a population can only arise from random variations and natural selections of the population, and definitely not from individual learning. Thus, it is important to understand that variations take place at the genotype level, while learning is based on actual performance in a given environment, namely at the phenotypic level. These principles should be carefully respected in evolutionary approaches and those mechanisms imitating their operation.

3.3   Evolutionary Computing

When trying to identify the most powerful natural problem solvers, which would be desired to be imitated and/or exploited in engineering approaches, and more specifically in the design of wireless networks, these two are the most characteristic ones:

1. the human brain

2. evolutionary process

The latter is essentially created by the human brain (and other brains too, to certain degrees) and is sometimes also referred to as cognition or cognitive process. Designing solvers based on the first item is part of the field of neuro-computing. The second, however, forms the basis for evolutionary computing.

The evolutionary computing field provides a proper framework for developing mechanisms and algorithms applicable to a wide range of problems (as the cognitive operations of nature), which do not need much tailoring for specific problems and deliver good (but not necessarily optimal) solutions within acceptable time scales. The latter, namely sub-optimality, which essentially satisfies various time and possibly resource constraints, is the essence of cognition, which is rather clearly observed in human behavior and other natural processes.

In automated problem solving, one may observe three main components, namely *inputs*, *internal models* receiving the inputs and producing the third component, namely *outputs*. Knowledge of the model essentially means knowledge of the system, since given an input and the model, the output may be computed. In that rather generic consideration of systems as input-model-output, three major categories may be identified in the literature, depending on the knowledge availability of each component. The three classes of systems are shown in Figure 3.1.

Figure 3.1(a) describes an *optimization problem* type, where the model and the desired output are known and the objective is to find the required input or inputs ensuring the desired output. Optimality is one of the desired

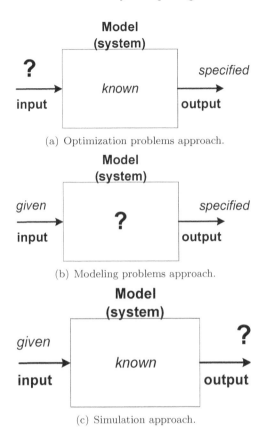

(a) Optimization problems approach.

(b) Modeling problems approach.

(c) Simulation approach.

Figure 3.1: Problem solving approaches.

properties that the system should bear and in communications terms this is usually in the form of minimum resource consumption or other paid cost, or maximum obtained outcome (benefit).

Figure 3.1(b) depicts a *modeling* or *system identification* problem type, where now the input/output of the system are known and the objective is to obtain a model that provides the correct output for each known input. In such cases, typically prior input-output combinations history is exploited for seeking a suitable model, which is aimed to be used as a predictor that will provide the output for future received inputs.

Finally, Figure 3.1(c) presents a *simulation problem* type of system, where the model and some inputs are known, while the objective is to compute the corresponding outputs. Simulation is used extensively, especially in network engineering in communications, as a cheap means to study the performance of developed network infrastructure and for developing, testing, and validating new concepts and protocol mechanisms.

Evolutionary computing has been applied and has potential applications in all three of the above classes of automated problem solving systems. For instance, the ant-colony optimization technique, which has been inspired by the observation of ant colonies, has been successfully applied in various diverse applications, such as for reverse optimization of car airfoils in mechanical engineering [53]. Evolutionary computing has also been applied in modeling problems, as in applications where available data, e.g., socio- and geo-location data, were required to be matched to the profiles of users/clients for mobile operators, financial institutions, etc. Such models are highly desired for their predictive powers and their applications in critical decision-making. Another characteristic case of modeling approaches are the design of classifiers, i.e., software systems that perform filtering and sorting of rules, trends, behaviors, etc. and have tremendous impact in the successfulness of applications in stock markets, marketing, trading, etc. Finally, evolutionary computing has been identified in the simulation domain as well, especially in finance and communications networks. One characteristic example is agent-based computation, where distributed software agents are employed for modeling the behavior of financial players or network nodes, and obtain performance indications and/or validation for developed schemes. In such cases, the outcomes need to be explored in a very cautionary fashion, by taking into account the specific assumptions of the simulation study and the conditions of the actual environment.

Sometimes, evolution may be considered as an adaptation process rather than an optimizer. In this case, the fitness does not correspond to an objective function, but rather as an expression of environmental or operational requirements. The populations strive to adapt to the imposed requirements, hence the notion of evolutionary adaptation. It is key to always keep in mind that several of the components of this process, which will be analyzed in more detail in the following sections, are stochastic.

3.3.1   Components of Evolutionary Algorithms

Many different variants of evolutionary algorithms exist. However, a common underlying concept that all these algorithms share is that given a population of individuals (solutions), the environment leads to natural selection (survival of the fittest), leading in turn in the rise of the overall fitness of the yielded population. In general, given a function to be maximized (minimization can usually be treated as a maximization of the inverse) a random set of candidate solutions can be created from the functions' domain. The quality function is applied as a fitness measure and the higher it is, the better for the search procedure. Some of the better candidates are chosen according to the fitness function, in order to seed the next generation by applying either recombination or mutation. The first combines features of two or more of the selected candidates (parents) and results in one or more new candidates (children). Mutation is applied to a single candidate and results again in a single new candidate. The set of new candidates (offspring) compete based on their fitness values and the whole process repeats until a solution (candidate set) with sufficient cumulative fitness (quality) value is achieved.

In general the process described above is based on two driving forces:

1. variation operator (in this case recombination and mutation)

2. selection

The first creates the required diversity of the trial-and-error search process, which essentially creates novelty in the considered solution set. The second acts as a quality pusher, forcing the selection of individuals (solutions) of better quality. The combined application of the above two forces leads to improved fitness in consecutive populations.

A general sketch of the modules and their operation of an evolutionary algorithm in pseudocode is given in the following Algorithm 1 and depicts the potential components that an evolutionary algorithm may consist of.

Input: *population*
Output: new *population*
begin
    INITIALIZE population with random candidate solutions;
    EVALUATE each candidate;
    while TERMINATION CONDITION *not satisfied* **do**
        SELECT parents;
        RECOMBINE pairs of *parents*;
        MUTATE the resulting *offspring*;
        EVALUATE new *candidates*;
        SELECT individuals for the next generation;
    end
end

Algorithm 1: General mechanism of an evolutionary algorithm (pseudocode).

It should be clearly noted that several of the components depicted in Algorithm 1 are stochastic, in order to ensure the better and more efficient exploration of the corresponding search space. However, this is also the reason that such approaches are sub-optimal, since the whole space is not systematically covered, and thus, the potentially uniquely optimal solution may be lost by the stochastic search strategy.

From the above description, it becomes evident that evolutionary algorithms employ a generate-and-test philosophy (we also referred to that previously as trial-and-error), where random outcomes are produced and with greater probability the fittest among these survive to the next round. The evaluation of fitness function is in fact a heuristic approach to estimate the quality (fit to our desires) of a solution candidate in the population. The search in the surviving population is driven by the variation and selection operators.

The most important components of an evolutionary algorithm can be identified from the pseudocode table of Algorithm 1, and they are provided in the following list:

1. Representation

2. Evaluation (fitness) function

3. Population

4. Parent selection

5. Variation operators, recombination and mutation

6. Survivor selection (replacement)

In addition to the definition of the above components, an initialization procedure(s) and termination condition(s) should be defined for each algorithm to operate properly and according to the specific application framework it is intended for. The whole process is schematically depicted in Figure 3.2.

3.3.2   Representation

Representation is the first step for defining an evolutionary algorithm and its objective is to link the real world with the "evolutionary world." Objects forming candidate solutions are referred to as phenotypes and their representation as genotypes, in accordance with the previous discussion on the genetics analogy. The representation part accounts for specifying a mapping from the phenotypes onto the set of genotypes, representing the original phenotypes. An example of that in the mathematical programming domain would involve

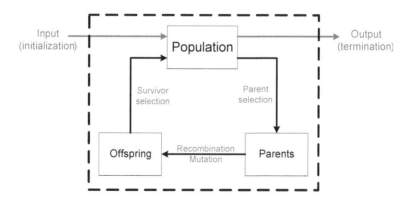

Figure 3.2: Schematic operation of a generic evolutionary algorithm.

integers as the phenotypes and their binary codes as their genotype representations. The terms *candidate solution* and *chromosome* have also been used extensively to denote the phenotype and genotype individuals, respectively.

The whole process in an evolutionary algorithm takes place in the genotype world, where an optimal (or suboptimal) solution is discovered and then a reverse process of translating the obtained genotype-domain solution to its phenotype-domain is required. The representation can be used in another sense than the above sense of "encoding" described already. In that second sense, it describes the data structure employed for the genotype in general.

3.3.3 Fitness Function

The role of the evaluation function, or *fitness function* as it is most commonly known in evolutionary computing literature, is to represent the adaptations required for the representation of reality in the evolutionary computing world. It facilitates the basis on which improvement takes place, and therefore, it can also enable selection leading eventually to evolution. Fundamentally, it defines the generalized notion of improvement in the evolutionary process, and mathematically, it is a function that assigns a quality measure to the genotypes defined from the selected representation. In some cases, such fitness function may be required to be minimized within a specific problem framework, but this should not be confused with the implied notion of improvement that the 'fitness' term bears. It should be treated simply as an objective function to be optimized in the sense dictated by the specific application framework applied into. In many cases, the fitness function may be defined as a functional or transformation of the actual objective function optimized in a mathematical program within the framework of the studied problem. However, in any case, the fitness is employed to assign a measure of improvement onto the candidate solutions (genotypes), based on which election of the the most superior will take place with higher probability, eventually yielding an overall more superior (fittest) candidate solution set.

3.3.4 Population

The *population* contains the representation of potential solutions, and as such, it is a multiset of genotypes, in the sense that multiple copies of an element are possible in this set. It forms the basis of evolution in that, while the individuals remain static and do not change, it is the whole population that changes and thus through population evolution, evolution of the complete system follows. Representations of the populations can be in one of multiple forms, e.g., enumeration, specifying population sizes, etc. However, in many cases, populations are equipped with an additional measurable property, e.g., distance or neighboring relation between the elements of each population. This additional structure may also take part in the selection process, as part of the fitness evaluation. Contrary to variation operators, which will be analyzed shortly, parent selection (explained right in the next subsection) is applied to the whole population, rather than on individuals in the population. Furthermore, in most of the evolutionary algorithm applications, the population size remains constant during each evolutionary step. Finally, another important aspect of a population is its diversity. It is defined as the number of different solutions that the population may obtain. In that sense, no single successful measure of population diversity exists, and thus, several different statistical measures, such as the entropy, for example, are usually employed for this purpose, typically adapted with respect to the specific application framework that the evolutionary algorithms are going to be used for.

3.3.5 Parent Selection

Parent selection aims at distinguishing among individuals based on their quality, and thus, allows the better individuals to become parents of the next generation. An individual of the population is called a parent if it has been selected to undergo one of the variations, which will be analyzed in detail in the next subsection. Parent selection is usually probabilistic and in conjunction with survivor selection, they both aim at driving the cumulative quality of a population at higher scales. This means that higher-quality individuals are preferred and should receive a higher chance to become parents. However, in order to maintain the trial-and-error character of an evolutionary algorithm for successive populations, low-quality (fitness) individuals should have some small but positive chance of becoming parents. Otherwise, the whole search process could get stuck in a local optimum of the search space.

3.3.6 Variation Operators: Recombination and Mutation

Variation operators create new individuals, the latter corresponding to generating new candidate solutions in the phenotype space. Based on the number of inputs they admit, variation operators are segregated in *recombination* and *mutations*.

Recombination (sometimes referred to as crossover) is a binary variation operator, i.e., it admits two individuals (genotypes) as inputs and produces one or two offspring genotypes. It is a stochastic operator and assumes different operations in various evolutionary computing paradigms. Recombination operators with more than two input genotypes are mathematically possible, but rarely employed, due to the lack of biological equivalence, even though they seem to have positive effects on evolution [9]. The principle behind recombination is to mate two individuals with different, but desired features and produce an offspring that combines those features.

On the other hand, a unary operator that admits only one genotype input is called mutation. It yields a modified mutant, called child or offspring. Mutations are stochastic operators and the offspring depends on the outcomes of a set of random choices, which, however, are unbiased. As with recombination, mutation may have different operations in different evolutionary computing paradigms.

Variation operators form the evolutionary implementation of the elementary steps within a search space. Generating children is essentially equivalent to stepping to new points in the search space. In that sense, mutation can also guarantee that the search space is connected (convex[2] in such cases). Also, variation operators are representation dependent, and they need to be explicitly defined in each case.

3.3.7 Survivor Selection

The role of survivor selection mechanisms is to distinguish among the individuals of successive populations based on their fitness (quality). Intuitively, it might seem similar to parent selection, but it is applied in a different stage of the evolutionary cycle, namely following the generation of offspring. Survivor selection is employed in order to maintain the size of the population in the original level, where, however, we want to ensure that the fittest offspring make it in the surviving population compared to the rest of the offspring. In addition to fitness selection of the offspring to make it to the next generation, sometimes the age (in terms of successive generations) of an individual is also taken into account.

Contrary to parent selection, survivor selection, which is also sometimes called replacement as well, is mostly of a deterministic nature. Classification and selection of the fittest individuals is often employed to implement replacement. The concept behind such deterministic survivor selection is that some randomness has already been incorporated in parent selection, and if another randomness level has been implemented, especially in the survivor selection, then the information conveyed by the fitness function, namely quality of individuals, would be canceled out by the random selection of survivors. Thus, the stochastic search is implemented in the parent selection step, and the effect of the offspring generation and surviving is implemented through the fitness function and deterministic survival selection.

3.3.8 Initialization and Termination Conditions

The initialization of evolutionary algorithms is usually kept simple, mostly through a simple randomly generated population of individuals. Alternative, problem-specific heuristics for generating the first population can also be devised at the cost of additional computation cost. In such cases, it is a matter of tradeoff balancing that depends on each specific problem that determines whether the added complexity of generating a more educated initial popula-

---

[2]For more details in convex search space and optimization, the interested reader is referred to [27] and [39].

tion would eventually yield significant benefit in the overall performance of the evolutionary algorithm.

In general there are two types of stopping conditions for evolutionary algorithms. The first is based on a predefined and desired level of the objective function, which once reached by the cumulative fitness value of the population signifies the stopping of the algorithm. This is the case in problems and environments, where the desired optimal fitness level is known a priori, or it can be computed rather easily. In the second type of stopping conditions, the optimum/desired fitness level might not be reached after a large number of iterations, or not at all. Alternative conditions should be set, which stop the evolutionary cycle with certainty. Some of the candidates are the following:

1. Defining a maximum number of evolutionary iterations.

2. Defining a maximum number of computations.

3. The fitness improvement remains restricted by a given threshold for a given time period (expressed in evolutionary iterations).

4. The population diversity drops under a given threshold.

In many cases, a combination (disjunction) of the above two termination conditions is employed. Namely, the evolutionary algorithm terminates "if the optimum fitness level is hit" *or* "some specified condition is satisfied."

3.3.9   Operation of Evolutionary Algorithm

In traditional optimization approaches (continuous) the search follows monotone directions, until a global optimum is reached, or until the search is trapped in a local optimum. Evolutionary algorithms operate in a different manner. In the early stages, the individuals are randomly spread over the search space. After some evolutionary iterations the population concentrates around local optimums. By the end of the evolutionary algorithm the population has concentrated over very few optima, some of which can be local and some global optima.

In principle, it is possible that the population concentrates around the wrong optimum (potentially a local one yielding a suboptimal solution eventually). These search phases of an evolutionary algorithm are referred to as *exploration* (when new individuals are generated in untested regions of the search space) and *exploitation* (in the event of concentration of the search in the vicinity of known good solutions). The evolutionary search process often comes down to balancing the tradeoff between exploration and exploitation, where too much exploration leads to inefficient search and slow convergence, while too much of the second leads to very focused searches that lead to what is referred to as *premature convergence*, that is, losing population diversity too quickly, which in turn leads to trapping in a local optimum. The latter is a considerable risk that all evolutionary algorithms face and should be taken into account in the design of such approaches.

Another distinctive feature of evolutionary algorithms, and many other approaches as well, is the *anytime property*. This feature describes the fact that an evolutionary algorithm (and the other approaches too with the same property) are capable of yielding a solution at anytime the search is stopped, even if this solution is suboptimal. If the search is stopped in the very early parts of the search, the solution may well be far away from the optimal (or acceptable suboptimal solutions). However, in any case a solution will be available. This property is characteristic of approaches that work under the notion of iterative improvement. Evolutionary algorithms fall under this category. Bearing the above property has some interesting consequences for evolutionary algorithms. Since from the anytime property a solution is always available and given the typical progress curve of such approaches, shown in a generic form in Figure 3.3, it could be concluded that the initialization process of an evolutionary algorithm does not have to be very complex and a simple one will suffice, since within a relatively small number of evolutionary iterations, the initial random population will quickly improve its fitness, while saving the additional cost of developing and obtaining an educated initial population. A similar consequence involves the termination condition of evolutionary algorithms. As observed in Figure 3.3, beyond a point in time, additional evolutionary iterations asymptotically improve the fitness function. This may be exploited so that significant resources are saved, once an acceptable quality solution is attained (and at the same time preventing long runs that have no significant improvement in the overall performance).

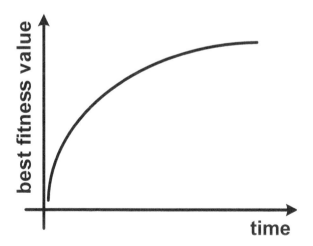

Figure 3.3: Typical form of the evolution progress of evolutionary algorithms (progress of the best fitness function).

Another aspect of evolutionary algorithms characterizing them cumulatively is that most of them are well-rounded and fit many types of problems and their variations. In most of the cases, they can act as robust problem solvers, since they provide a roughly good performance over a wide range of problems, while not requiring significant implementation complexity, effort, and operation. On the other hand, specifically developed algorithms perform much better within the framework of the problem they were designed for, or for a specific variation of the problem, where they exploit special structures of the specific search space or of the special structure of a problem instance. However, as the analysis deviates from the specific problem type to other variations, or other similar problems, problem-specific algorithms quickly lose performance benefits yielded. Consequently, facing a problem, there are essentially three available options, namely an evolutionary algorithm, a problem-specific one, and one based completely on random search. Each of them stands in a different position regarding the analyzed tradeoff between exploration and exploitation. The evolutionary approach is a good compromise for covering most aspects of a problem at an acceptable level.

Regarding evolutionary algorithms and their provenly optimal counterpart algorithms, it is widely recognized that in general, irrespective of the progress of computational power, most real-world problems are reduced to well-known abstract forms, for which the number of potential solutions grows exponentially with the number of considered variables, when provenly optimal algorithms are employed. This means that beyond a certain problem size (which varies for different problems), the search for provably optimal solutions cannot be attained and alternative (sometimes even heuristic approaches) are required for obtaining good and acceptable solutions. Apart from the exact methods, which even for some explicit boxing methods that arrange solutions in tree structures and intelligently prune several branches guaranteed not to provide good solutions (but still do not significantly reduce computational cost), a class of search methods (heuristics), such as simulated annealing, are guaranteed to find the optimal solution discovered until that point of their search, but not necessarily the optimal one. Other categories of algorithms that perform local search actually search only within restricted neighborhoods of solutions from an already obtained solution point, but these too potentially lead to local optimum solutions, rather than provenly global optimum ones. However, they are good in quickly providing solutions with usually acceptable solution quality (fitness).

Evolutionary algorithms distinctively distinguish from previous local search algorithms due to the use of the population notion, which allows maintaining a diverse set of search points, which in turn not only enables escaping from local optimum points, but also provides a means to tackle the large and even discontinuous search spaces. This contrasts with the globally uniform distribution of purely random search or the locally uniform distribution of other stochastic algorithms, e.g., simulated annealing,[3] and in several cases, it presents opportunities for achieving performance and efficiency.

3.4 Evolutionary Computing Approaches

In this section, we present several evolutionary approaches/algorithms that have been shown to apply in several problem types and yield convenient solution results.

3.4.1 Genetic Algorithms

Perhaps the most widely known type of evolutionary approach is the *genetic algorithm*, which as its name indicates, is inspired by the genetics field. A general feature of this approach is that there is actually no single definitive genetic algorithm, but rather designers create algorithms from a suite of operators in order to accommodate their particular needs and the requirements of each problem.

Genetic algorithms were first introduced by Holland [87] for studying adaptive behaviors in natural but mainly in artificial systems. In general, they have been customarily considered as function optimization methods, which, however, is not always the case.

In the following, we describe the various components of genetic algorithms as identified in previous sections. We start with the representation of candidate solutions, then proceed with variation operations, such as mutation, followed by parent selection and survivor selection.

Representation of Individuals

The representation defines the genotype and the mapping from genotype to phenotype. Choosing the right representation is a key factor of the performance and efficiency of an evolutionary algorithm and thus of a genetic algorithm as well. For this reason, a number of alternative representations for genetic algorithms will be presented in the sequel, and in most cases, it comes down to the complete set of problem parameters for deciding which representation is more appropriate. It frequently turns out in practice that using mixed representations is a more suitable way of describing and manipulating a problem search process.

---

[3]Simulated annealing (SA) is a generic probabilistic meta-heuristic for the global optimization problem of locating a good approximation to the global optimum of a given function in a large search space. It is often used when the search space is discrete (e.g., assignment of white space channels among cognitive radio secondary users). For certain problems, simulated annealing may be more efficient than exhaustive enumeration provided that the goal is merely to find an acceptably good solution in a fixed amount of time, rather than the best possible solution. SA implements the notion of slow cooling inspired by annealing in metallurgy, as a slow decrease in the probability of accepting worse solutions as it explores the solution space. Accepting worse solutions is a fundamental property of meta-heuristics because it allows for a more extensive search for the optimal solution. The more interested reader can refer to [28] and references therein for more details.

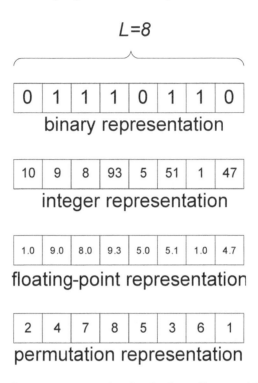

Figure 3.4: Representation of individuals in Genetic Algorithms.

One of the simplest representations, if not the simplest, is the binary (Figure 3.4), where the genotype consists of a bit string (string of binary digits). For some problems, especially when the decision variables are Boolean, the mapping to a binary genotype is natural, since the phenotype is in a similar form. However, in many cases, bit strings are employed to encode non-binary information. An important factor is the selection of bit string length employed. The encoding size should allow all possible bit strings to denote a valid solution to the given problem and vice versa, all possible solutions to be represented. Finally, the interpretation of the obtained genotype to an appropriate phenotype should be ensured in an accurate manner.

Integer representations (Figure 3.4) are an alternative to binary ones, especially in problems where the objective is to find optimal values for variables so that they all take integer values. Apart from the obvious value range restrictions, integer representation might also have ordinal attributes. However, in some cases integer representations might be suitable, but do not necessarily have ordinal properties. Such cases involve, for instance, variables representing azimuthal information, where integers are suitable for representation, but do not have a natural ordering as the set of integers.

A sensible and usual representation is that of real numbers (floating point representation), which is capable of accommodating a broad range of application frameworks. This representation is usually more applicable to cases where the gene values are derived from a continuous domain/distribution, rather than from a discrete one (Figure 3.4). In these cases, machine precision could be a limiting factor, or a factor to take seriously into account, in order to achieve/maintain an acceptable or desired level of accuracy.

In the event that the objective of the genetic algorithm is to decide the order in which the sequence of events should occur, or a similar decision, a permutation representation (Figure 3.4), based on a permutation of a set of integers, is more appropriate than all the above. The specific representation is required to ensure that each possible value occurs exactly once in each candidate solution (genotype). Especially regarding permutation representation, two types of permutations should be distinguished. The first is where the order of permutated integers matters, and the second is where the adjacency of the represented individuals/entities matters. In the second type of permutations, even though the sequence of elements of the permutation matters, the initial point does not, so that only the consistency of the sequence should be ensured, but not necessarily the specific traversing.

Examples of all four representation types for genetic algorithms are cumulatively shown in Figure 3.4.

Mutation

As already mentioned, mutation refers cumulatively to the set of operators that have input a single parent and provide as output a single child by performing some randomization to the representation of the corresponding representation (genotype) of the individual. The form of the mutation depends greatly on the underlying representation employed. In addition, notable impact on the behavior of the algorithm could be caused by the context of the parameters (mutation rates) associated with mutation. Since mutation depends on the representation, in the following, we discuss mutation operators for the representations mentioned above.

For binary representations, the most frequently used mutation operations treat each bit in the representation string separately and allows it to flip independently with a specified probability p. Clearly, the choice of p determines the suitability of the operator for each application framework that the genetic algorithm is used for. Thus, the most suitable choice of p is usually determined in a training phase for each genetic algorithm, where this and another series of tunable parameters are fine-tuned, given prototype behaviors and desired objectives. Due to the independence of each bit flip, the number of bits changed in each mutation is not fixed, but on average $L \times p$ bits are flipped, where L is the encoding length L selected. In this case, the probability p is the mutation rate of the operator.

The integer representation has available two main types of mutation operators, both mutating each bit (gene) independently with mutation rate p. The first is *random resetting*, where bit flipping is extended to randomly choosing a new value from the set of permissible values for each gene. The second is the *creep mutation*, where a small positive or negative value is added to each gene with probability p'. The first is more appropriate for encodings with cardinal attributes, while the second for representations with ordinal attributes. The small values added in each gene for the creep mutation are usually selected from symmetric distributions with mean around zero and small variance, so that the changes are likely to be small (sometimes referred to as perturbations). Thus, creep mutation requires additional parameters, in order to control the perturbation distribution, in addition to the probability of perturbing each gene, compared to the random resetting mutation.

For floating-point representations the gene values are derived from continuous distributions and thus, it is common to change the value of each gene randomly within its corresponding domain, denoted in the following by $[L_i, U_i]$. Two major types of floating-point related mutations maybe identified, *uniform mutation* and *nonuniform with fixed distribution mutation*. The first essentially extends random resetting, where now the values for the genes are drawn randomly and uniformly from $[L_i, U_i]$. It is usually used in conjunction to position-wise mutation probability for selecting the gene to mutate. Nonuniform mutation, on the other hand, extends creep mutation and it is designed so that a perturbation is added to the current value of a gene. The fixed distribution is usually a Gaussian with zero mean and specified standard deviation and then curtailing the yielded value to the specified range $[L_i, U_i]$ if needed. In this case, most of the changes will be small (perturbations), but nonzero probability for greater changes remains as in the creep mutation. Usually, this mutation is applied per each separate gene and the only parameters used are those controlling the fixed distribution employed (mean and standard deviation in the Gaussian distribution case).

In permutation representations and in order to maintain the consistency of the encoding string employed, the mutation parameter is meant in the sense of the probability that the string undergoes mutation, rather than that a single gene of the string is altered. The most common permutation oriented mutation operators are the *swap mutation*, where two genes in a string are randomly selected and their values are swapped, the *insert mutation*, where two genes are randomly selected and the one is moved next to the other by shuffling along the other genes that were originally between them, the *scramble mutation*, where the entire string or some randomly selected subsets of the genes have their positions scrambled, and *inversion mutation*, where two positions in the string are randomly selected and the order in which the values of the genes between the two selected positions appear is reversed. This can be very useful for solving demanding problems, since this inversion of randomly chosen substrings is the smallest change that can be made to, e.g., an adjacency-based problem, which uses permutation representation, such as solving the TSP problem with the 2-opt search heuristic [106].

Figure 3.5 and Figure 3.6 cumulatively depict the mutation operators for

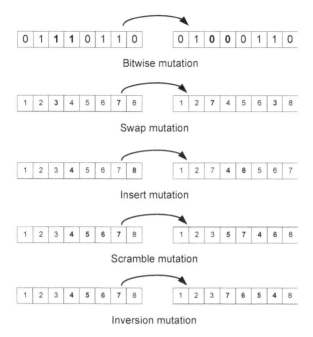

Figure 3.5: Mutation operator examples in Genetic Algorithms.

Genetic Algorithms and their classification based on the representation employed in each case.

Recombination (Crossover)

Recombination is considered one of the most important features of a genetic algorithm, and it combines information from two or more parents to produce a new individual solution. Compared to recombination, mutation as explained above is considered secondary for creating diversity, and thus, significantly more research has been devoted to the development of suitable recombination methods.

Recombination is sometimes referred to as crossover as well. The reason lies in the main concept implemented by recombination. A crossover rate p_c is defined, which lies in the range $[0.5, 1]$, and two parents are selected. Then a random variable in $[0, 1)$ is drawn and compared to p_c. If the value is lower, two offspring are created through recombination of the two parents, otherwise the parents remain intact and are yielded as the offspring. In contrast to the mutation probability p mentioned above, which also controls how parts of the chromosome are perturbed independently, the recombination probability determines the probability that a chosen pair of parents will undergo the recombination operator.

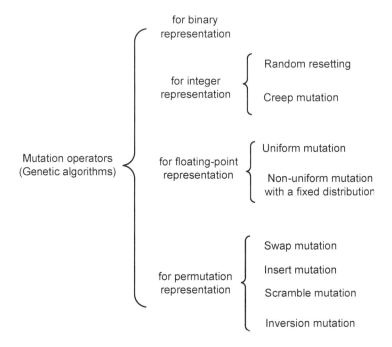

Figure 3.6: Mutation operators in Genetic Algorithms.

The recombination operators for binary and integer representation can have the same implementation. It is also possible that some of the operations on the values of the genes, e.g., in the event of integer representations, might yield values not in the original set, namely non-integers in the case of integer representations. Thus, recombination operators lead oftentimes to the sort of "blending" of gene values as the one described above.

The three main types of crossover operators for binary and integer representations are the *one-point crossover*, the *N-point crossover*, and the *uniform crossover*. One-point recombination works by choosing a random number in the range $[0, L - 1]$, L being the length of the string representing the encoding, and then splitting both parents at this point (position) and creating two children by exchanging the constituting parts. For binary representations, this seems straightforward to implement, while for integer representations the same procedure may be applied, where now in place of each bit one has integer values to work on. In the N-point crossover, the one-point crossover is extended to brake the representation in more than two substrings/subsequences. In this case, N randomly selected crossover points are selected in the range $[0, L - 1]$, and the offspring are created by taking alternative substrings/subsequences from the two parents, i.e., combining segments 1, 3, 5, etc. of the first with 2, 4, 6, etc. of the second and vice-versa. In the uniform crossover, each gene is treated independently and makes a random choice as to which parent it should be inherited from. To achieve this, ℓ random variables are drawn from the $[0, 1]$ distribution and for each position i in the representation string, the i-th random variable is compared with a parameter p (usually $p = 0.5$). If the random variable is lower than p the gene is inherited from the first parent, otherwise from the second. The second offspring is created using the complementary process.

From the above discussion, it might be natural to believe that recombination in the sense discussed already works by randomly mixing parts of the parents; however, N-point crossover has an inherent bias in that it tends to keep together genes that are located close to each other in their original representation (since whole chunks of the original string are inherited in the offspring). In addition, the parity (even or odd) of the string length might create strong biases against keeping together combinations of genes originally located in opposite ends of the encoding string. Such effects are known as *positional bias*. On the contrary, uniform crossover does not exhibit positional bias, but rather has another tendency known as *distributional bias*, where approximately 50% of the genes from each parent are transmitted to the offspring and at the same time there is a negative tendency in transmitting to the offspring a large number of co-adapted genes originating in one of the two parents.

In general, understanding the types of biases involved in each recombination operator is invaluable for developing algorithms for specific problems, particularly when dependencies or known patterns emerge in the representation.

For floating-point representations, the recombination operation becomes more complicated. There are two general types of crossover for floating-point representations. The first is an operator analogous to the one for binary representations, where now each gene value is a floating-point value, rather than a single bit. This process is known as *discrete recombination*, but suffers the drawback that it does not insert new values in the crossover (and the same holds for all the crossover recombinations presented above), leaving only mutation to insert such diversity. The second type of floating-point recombination, termed as *intermediate* or *arithmetic crossover*, creates for each gene position, a new value in the offspring that lies between those of the parents, as an arithmetic combination of the bounds set by the values of the gene parents. Formally, if x_i and y_i are the values of the parents for position i, then the offspring value is $z_i = \alpha x_i + (1 - \alpha)y_i$ for some $\alpha \in [0, 1]$. In this way, the recombination creates new gene values, but as a result of the averaging process the range of the values in the new population is reduced compared to that of the original population.

Three types of arithmetic recombination exist, in all of which the choice of the parameter α is made at random in $[0, 1]$ in theory, but in practice a fixed value of 0.5 is employed (*uniform arithmetic recombination*). In the first, denoted by *simple recombination*, a recombination point k is picked for the first child and the first k floats of the first parent are copied to the first child. The rest is the arithmetic average of both parents. The second child is created analogously. In the second, namely *single arithmetic recombination*, a random position k is picked and at this position, the arithmetic mean of the two parents is taken, while the other positions are determined from the parents. Finally, in the third type, *whole arithmetic recombination*, the weighted sum of the two parent values are taken for each gene.

Apart from mutation and recombination that utilize one and two parents respectively, it is straightforward to consider operators that use more parents. The resulting *multi-parent recombination* operators are simple to define and they are implemented by extending the other operators. This concept deviates from the traditional model observed in nature that inspired genetic algorithms, by moving to more peculiar 'reproduction' schemes. The latter is potentially a perspective for further improving recombination from a technical point of view. These operators can be classified on the basis of the main idea employed for combining parental information. More specifically, the following list presents some of these candidate operator categories:

1. based on gene frequencies (extends uniform crossover)

2. based on segmentation and recombination of parents (extends N-point crossover)

3. based on numerical operations on real-valued genes (generalizes arithmetic recombination operators)

However, in general, it is not definite that increasing the number of combined parents increases performance and efficiency of evolutionary algorithms. More systematic research is required in this case, even though several works [159] have exhibited potential benefits of combining more than two parents to obtain offspring candidate solutions in evolutionary algorithms.

Recombination for permutation-based representations appears to be more complicated, since the permutation property needs to be maintained even for this operator. Some specialized recombination operators have been developed for this case with the objective to transmit as much information contained in the parents as possible. The first is *partially mapped crossover*, which works as follows:

1. Two crossover points are chosen at random and the segment between them from the first parent is copied into the first offspring.

2. Starting from the first crossover point one searches for the elements of the second parent that have not been copied.

3. For each of these, we compare it with the element that was copied in the offspring in the specific place from the first parent.

4. The element copied in the offspring is placed in the corresponding position of the second parent.

5. If the place occupied in the second parent has already been filled in the offspring by an element k, then the element from the first parent is put in the position occupied by element k in the second parent.

6. Having dealt with the elements from the crossover segment, the rest of the offspring can be filled from the second parent, and the second child is created analogously with the two parental roles exchanged.

One of the desired properties of recombination is that of "respect," where any information carried in both parents should also be present in the offspring. This is true for the recombination operators for binary and integer representations and for discrete recombination for floating-point representations. However, this is not true for the partially mapped crossover. For this reason, several alternative operators have been designed for adjacency-based permutation problems. One such candidate is the *edge crossover*, which is based on the idea that an offspring should be created as far as possible using only edges[4] that are present in one or more parents. The operator works by building edge tables (adjacency lists), which for each element contain all other elements linked to the specific element in the two parents. The second, *order crossover*, begins in a similar manner to partially mapped crossover, by copying a randomly selected segment of the first parent to the offspring. However, its intention is to transmit information about relative order from the second parent. Thus,

1. Starting from the second crossover point in the second parent, the remaining unused numbers are copied into the first child in the order that they appear in the second parent, by wrapping around at the end of the list.

2. The second offspring is created in an analogous manner, with the parent roles inverted.

Finally, the third, denoted by *cycle crossover*, aims at preserving as much information as possible about the absolute position in which elements occur. The operator works by selecting alternate cycles from each parent, after the elements have been divided into cycles. A cycle is a subset of elements that has the property that each element always occurs paired with another element of the same cycle when the two parents are aligned. A more detailed process for constructing cycles, as well as details on implementing crossover operators for the permutation representation, may be found in [59].

Figure 3.7 presents recombination operators cumulatively for Genetic Algorithms, classified according to their employed representation of individuals. The figure does not contain the multi-parent operators, which could also be used in given circumstances.

---

[4]Here the notion of edge denotes elements linked together in the parents of offspring produced.

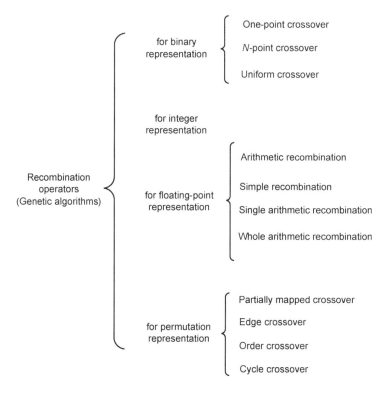

Figure 3.7: Recombination operators in Genetic Algorithms.

Population Model

The previous components of the genetic algorithms were focused on the representation and variation of candidate solutions, in order to provide a desired degree of diversity in the solution search process. Once successful in the previously described objective, the next one is to concentrate on the survival of these individuals and ensure they take part in the coming populations based on their fitness values.

In general, there are two models for genetic algorithm population, namely the *generational model* and the *steady-state model*. In the first, starting with a population of size μ, $\lambda = \mu$ offspring are created by the application of various operators and the population is evaluated based on the fitness value. The whole population is replaced by its offspring in each iteration, and the new population is referred to as "next generation." On the other hand, in the steady-state model the entire population is not changed at once, but only a percentage of it. In this case, $\lambda < \mu$ and λ old individuals are replaced by λ new offspring. The percentage of the population that is replaced is called the *generational gap* and equals λ/μ. Usually a generation gap $1/\mu$, with $\lambda = 1$, is employed in practical applications.

Parent Selection

In the description of the generic evolutionary cycle, there were two points of the cycle where fitness-based competition/selection took place. The first was in the mating process (parent selection) and the second in the selection of individuals to survive into the next generation (survivor selection). While the latter will be the topic of the following subsection, the parent selection process will be the focus of this one. The most frequently used parent selection approaches are described as follows.

The first one is the *fitness proportional selection*, where the probability that an individual is selected is proportional to the absolute fitness value of the individual to the cumulative absolute fitness value of the rest of the population. Even though simple, this approach is associated with some problems:

- Outstanding individuals take over the rest of the population very quickly (premature convergence).

- When all fitness values are close to each other, there is no *selection pressure*, i.e., selection becomes almost uniformly at random, which eventually means that once the worst individuals have disappeared quickly, performance improves only very slowly.

- The mechanism behaves differently on transposed instances of the same fitness function.

To avoid the last two problems mentioned above, windowing may be employed. Windowing is a process to keep track of the relative range of values a quantity might take. It maintains fitness differentials by subtracting from the fitness value a value that depends on the recent search history. A simple way to achieve this is simply to subtract the value of the least-fit member of the current population, or a running average over the last few generations, if the least-fit value fluctuates very rapidly.

The second is *ranking selection*, which preserves a constant selection pressure by sorting the populations according to the fitness value and then allocating selection probabilities to individuals according to their rank, rather than according to their fitness value. Clearly, the mapping from rank to selection probability can be arbitrary and incorporate different types of information desired for the selection strategy. The only constraint imposed is that the selection probabilities must sum to one, as expected. Usual mappings are linear and exponentially decreasing ones. In the first case, the selection pressure that can be applied is limited, while the exponential one induces higher selection pressure.[5]

In general, it would be desired to be possible to have the same selection likelihood for each individual in the population, so that the search space is potentially evenly covered. However, this is not the case in reality, due to the

---

[5]Here selection pressure refers to the selection probability. The higher the probability, the higher the selection pressure will be.

finite size of the population. The simplest way to achieve the desired sampling in a more efficient manner is denoted by *roulette wheel algorithm* and it works as follows. It assumes some ordering of the population, which defines a list of values, each of which corresponds to the population individuals and depends on the selection distribution fitness. This approach is equivalent to spinning a roulette wheel, where the sizes of the holes reflect the selection probabilities. *Stochastic universal sampling* is an alternative method preferred, especially whenever more than one sample is to be drawn. This is equivalent to making one spin of a wheel with equally spaced holes (as many as the population size), rather than multiple spins (as many as the population size) of a roulette wheel.

In cases of very large populations or distributed populations and operation, the above approaches are not suitable, since they rely on knowledge collected from the whole population. In such cases, user subjectivity accounts for the lack of a global quantitative measure that assigns fitness value to each member of the population. *Tournament selection* is an operator with this property, namely it does not require global knowledge of the population. Instead it relies on an ordering relation between any two individuals of the population. Tournament selection compares relative fitness rather than absolute and thus, it has the same properties as ranking schemes regarding transposition of fitness function and invariance translation. The selection probabilities in tournament selection depend mainly on:

- Rank of an individual in the population;

- The tournament size;

- The probability that the most fit member is selected;

- Whether individuals are chosen with or without replacement.

More details on the impact of the above parameters on the selection probabilities can be found in the more thorough treatment in [59]. Tournament selection has the drawback that the outcomes can show a high variance from the theoretical probability distribution of fitness (since tournament selection employs the user-subjective one). Despite that, it is perhaps the most widely used selection operator in contemporary genetic algorithms.

Survivor Selection

The main objective of survivor selection is to select among μ parents and λ offspring the μ individuals that will constitute the next generation, which is sometimes called replacement. Replacement mechanisms are usually differentiated on the basis of age or fitness of individuals and in the following, we present some of these replacement strategies.

In age-based replacement the fitness value is not taken into account, but rather only the number of iterations that the individual has survived. In this

Table 3.1: Summary of simple Genetic Algorithms.

Elements	Genetic Algorithms
Typical problems	function optimization
Representation	bit strings
Recombination	1-point crossover
Mutation	bit-flip
Parent selection	fitness proportional
Survivor selection	generational

way, each individual exists in the population for the same number of iterations. In genetic algorithms, a simple strategy where $\mu = \lambda$ the number of parents is the same as the number of offspring, each individual survives only one cycle, and the parents are discarded and replaced entirely by the offspring. The strategy can be implemented for overlapping populations $\lambda < \mu$ and single-individual ones as well. Each parent can be randomly selected for replacement, which, however, exhibits greater population variance, since in the random parent replacement it was likely to lose the best member of the population. For this reason, the random replacement strategy is not usually employed.

Fitness based replacement selection has been discussed in the previous section as well, such as fitness proportionate, tournament selection, and rank-based selection. Two other common mechanisms are *replacement worst* and *elitism*. In the first, the worst λ members of the population are selected for replacement. Even though this could lead to fast improvements, it could also lead to premature convergence, and thus, it is usually recommended for large populations, where the size will ensure that convergence time will be sufficient to include the desired diversity in the search process. In elitism, a trace of the currently fittest individual is maintained. If this member is chosen in the replacement group of individuals, and none of the replacement offspring has equal or better fitness, the traced individual is retained and one of the offspring is discarded in some random or more intelligent manner. Elitism is usually combined with age-based and stochastic fitness-based replacement, in order to prevent loss of the fittest individual in these approaches.

Table 3.1 summarizes the main features emerging in simple Genetic Algorithms (the basic flavor within the whole suite of Genetic Algorithms as presented above), and it can be compared with Table 3.2 summarizing the corresponding features for the rest of the evolutionary algorithms presented.

3.4.2   Evolutionary Strategy

Evolutionary strategies are characterized by another important feature of evolutionary computing, namely *self-adaptation*. Self-adaptation cumulatively characterizes the process that some parameters of an evolutionary algorithm can vary during a run of the algorithm (between successive iterations, but remaining fixed for a specific evolutionary iteration) in a specific manner. The varying parameters essentially co-evolve with the solutions. The perturbation parameter employed in Evolutionary Strategy (ES) for mutation is referred to as mutation step-size and it is rather important in the evolutionary cycle of evolutionary strategies.

Evolutionary programming strategies exhibit some typical characteristics, as shown in the following list:

- They are typically used for continuous parameter optimization, e.g., minimizing an n-dimensional function $\mathbb{R}^n \to \mathbb{R}$.

- Strong emphasis is put on mutation for creating offspring.

- Mutation is implemented by adding some random noise obtained from a Gaussian distribution.

- Mutation parameters vary for a run of the evolutionary algorithm.

The employed representation is real-valued vectors (since it is most frequently used for numerical optimization of continuous functions). Recombination is usually in the form of discrete or intermediary recombination and mutation is a Gaussian perturbation. Parent selection is uniformly random and survivor selection is (μ, λ) or $(\mu + \lambda)$.[6]

In representation for evolution strategy, the chromosomes consist of three parts, namely one with the object parameters $x_1, ..., x_n$ and another two with the strategy parameters, i.e., the mutation step sizes $\sigma_1, ...\sigma_{n_\sigma}$, and rotation angles $\alpha_1, ...\alpha_{n_\alpha}$. However, not every component of the above three is always present. The full representation vector is given by $< x_1, ..., x_n, \sigma_1, ...\sigma_{n_\sigma}, \alpha_1, ...\alpha_{n_\alpha} >$. In the simple case where self-adaptation is not utilized, the genotype space becomes identical to the phenotype $\mathbb{R}^n$ and no special encoding is required.

For mutation, the main mechanism is to change the value of parameters by adding random noise drawn from a normal distribution, such as $x'_i = x_i + N(0, \sigma')$. The key idea here and the actual implementation of the self-adaptation is that the mutation step-size σ is part of the chromosome and σ is also mutated into σ' so that the mutation step size co-evolves with the solution x. The net mutation effect is that $< x, \sigma > \to < x', \sigma' >$, where the order of update matters, requiring that the mutation step size evolves first and then the solution update follows. The rationale is that this order allows double evaluation of the net mutation $< x', \sigma' >$, where in the first x' is good if $f(x')$ and in the second, the σ' is good if the x' it created is good as well. By using a Gaussian distribution, small mutations are more likely than larger ones. Also, regarding the operation of varying mutation step sizes as explained above, it is implicitly assumed that under different circumstances different step sizes

---

[6]Survivor notation denotes that in the first only λ offspring out of μ parents are selected, while the second parameter denotes that λ offspring are selected from $\lambda + \mu$ parents.

will behave differently, and thus, some will be better than others. This turns self-adaptation to a mechanism adjusting the mutation strategy as the search of the evolutionary strategy proceeds. Assigning a separate mutation strategy to each individual, which coevolves with it, opens the possibility to learn and use a mutation operator suited for the local topology.

Different types of mutation can be employed, e.g., uncorrelated mutation with one step size, uncorrelated mutation with n step sizes, and correlated mutations. In the first, the same distribution is used to mutate each x_i, therefore there is only one strategy parameter σ associated with each individual. This parameter is mutated each time step by multiplying it by an exponential term e^Γ, where Γ is a random variable drawn each time from a normal distribution with zero mean and standard deviation τ. It is usually suggested that the employed standard deviation is not very close to zero, which is ensured by adopting specific boundary rules. Parameter τ is a kind of learning rate. The mutation of σ is of lognormal nature and the reason is that small modifications are ensured, while the standard deviations are sufficiently greater than zero and the mutation process remains neutral on average. When dimensions of the search space are desired to be treated differently, namely using different step sizes for different dimensions, uncorrelated mutation with n step sizes is applicable. The intuition behind this approach is that the fitness domain can be itself asymmetric, having different slopes in different directions along the various axes. In this case, each chromosome is extended with n step sizes, one for each dimension. The mechanism now varies for each dimension, but again a boundary rule is employed to prevent standard deviations of the mutation step sizes from approaching zero. The rule becomes:

$$\sigma'_i = \sigma_i e^{\tau' N(0,1)+\tau N(0,1)} \tag{3.1}$$

$$\chi'_i = \chi_i + \sigma_i N_i(0,1) \tag{3.2}$$

where the term $e^{\tau' N(0,1)}$ is a common mutation base that allows for an overall change of the mutability guaranteeing preservation of all degrees of freedom, while the coordinate-specific $e^{\tau N(0,1)}$ provides the flexibility to use different mutation strategies in different directions. Considering the geometry of the search space, the uncorrelated mutation with one step essentially corresponds to creating a circle around an individual, while the uncorrelated mutation an n-step size forms an ellipse. Generalizing this concept to ellipses with arbitrary orientations (rather than aligned with the coordinate system axes as in the uncorrelated mutation with n-step sizes) leads to correlated mutations, where the rotation of the ellipse takes place according to a covariance matrix C. The complete mutation mechanism is described by:

$$\sigma'_i = \sigma_i e^{\tau' N(0,1)+\tau N(0,1)} \tag{3.3}$$

$$\alpha'_j = \alpha_j + \beta N(0,1) \tag{3.4}$$

$$x' = x + N(0, C') \tag{3.5}$$

where β is a constant with proper dimensions, usually in the order of 5°.

Recombination on the other hand creates one child. It acts on each variable/position, either by averaging parental values (intermediary recombination), or by selecting one of the parental values according to a specified discipline (if parental values are randomly and uniformly selected, the recombination is called discrete). For more parents, recombination works similarly by using two selected parents to make a discipline and by selecting two parents anew for each position. In order to obtain λ offspring, recombination must be performed λ times.

Formally, the recombination can be given as:

$$z_i = \begin{cases} (x_i, y_i)/2 & \text{intermediary recombination} \\ x_i \text{ or } y_i \text{ chosen randomly} & \text{discrete recombination} \end{cases} \quad (3.6)$$

where x and y are the parent vectors and z is the child. A multi-parent variant of recombination is denoted by global recombination (while the original is referred to as local recombination) and more than two recombinants may be used in this case. Evolutionary strategies typically employ global recombination. However, typically discrete recombination is used for the objective variable part and intermediary recombination for the strategy parameters part.

Since parent selection is uniformly random, the selection is unbiased and thus, each member of the population can be potentially selected as a parent. Parents are randomly and uniformly selected from the population of μ individuals.

Survivor selection is applied after creating λ offspring from μ parents by mutation and recombination. If selection takes place from the offspring only, it is called (μ, λ) selection, otherwise, if offspring are selected from the union of parents and offspring it is called $(\mu + \lambda)$ selection. Both schemes are purely deterministic and based on rank rather than on absolute fitness value. The first is usually preferred in evolutionary strategies because it discards all parents and therefore, it leads to local optima in principle, without failing to follow the moving optimum of an on-fixed fitness function as $(\mu + \lambda)$ selection does. Also, it does not hinder the self-adaptation with respect to strategy parameters like $(\mu + \lambda)$ selection does.

The benefits of self-adaptation, which was first introduced in evolutionary strategies, have been shown not only for real-valued search spaces, but also for binary and integer ones. This inspired other types of evolutionary algorithms to adopt it as well. Unfortunately, there does not exist a firm theoretical validation on the effectiveness of self-adaptation. However, close match of the theoretical and experimental results seem to confirm the effectiveness of the approach. Both theoretical and experimental studies agree on the fact that the step sizes σ must decrease over time. The reason is that initially a large space has to be covered to locate the promising regions, while later smaller fine-tuning searches are required. Also, when the objective function changes, the current population needs to be re-evaluated and some individuals may now receive low fitness value, since they were adapted to the old objective. Self-adaptation is able to reset the step sizes after each change in the objective function and thus improve the search process accordingly. Some conditions that suit the employment of self-adaptation more are a generation of offspring surplus ($\lambda > \mu$), using (μ, λ) selection and the use of recombination on the strategy parameters as well.

3.4.3 Genetic Programming

Genetic programming is the most recent type of evolutionary algorithm. Apart from the basic representation difference with other evolutionary approaches (genetic programming uses trees as chromosomes), it also has a different application area. Genetic programming aims at seeking models with maximum fit, rather than finding some input (solution) that maximizes the obtained payoff, as provided by a specifically defined function (fitness function). Thus, genetic programming is essentially a specialization of the Genetic Algorithms (GA) option, where each individual is now a computer program. Consequently, genetic programming could be classified as a machine learning technique as well, which is used to optimize a population of computer programs, according to their fitness values, the latter determined by a program's ability to perform a given computational task.

The main applications of genetic programming are prediction, classification, and similar operations closely related to machine learning. Compared to other approaches, genetic programming competes with neural networks, since it is usually applied over large populations and it is typically characterized by slow evolution, namely it usually takes significantly large number of evolutionary iterations to converge to the desired solutions.

In genetic programming computer program populations evolve where each computer program is represented in memory as a tree structure. Such representations over trees have the advantage that they can be completely evaluated in a convenient recursive manner with logarithmic complexity. Every tree node has an operator function and every terminal (leaf) node has an operand, thus making mathematical expressions easy to evolve and evaluate recursively. Such expressions are formal notations for computer programs in manners very well documented in finite state machine theory. Such trees used to evaluate expressions in a given formal syntax are called parse trees. The expressions can be arithmetic, logical, or even codes for whole programs (e.g., in C programming language), as shown in the examples in Figure 3.8. Parse trees ensure that the expressions with the formal syntax can be properly interpreted in each case. From a technical perspective genetic programming is a variant of genetic algorithms operating on a specific and different than usual data structure, namely trees. Parse trees may be interpreted in various ways depending on the application context and employed representation.

For this reason, genetic programming favors programming languages and especially functional programming languages, like LISP, which naturally embody tree structures. At this point we note that LISP is the programming language most frequently employed in Artificial Intelligence (AI), where again

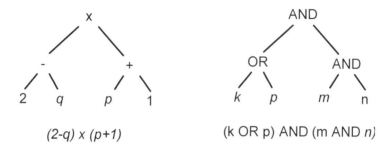

$(2-q) \times (p+1)$ (k OR p) AND (m AND *n*)

Figure 3.8: Examples of tree representations in Genetic Programming algorithms (arithmetic on the left — logical on the right).

evolutionary cognitive cycles like the one represented in the beginning of this chapter, and the one that will be presented in Chapter 6, play significant roles as many AI approaches strive to mimic evolutionary operations observed in natural phenomena.

The specification of individuals in genetic programming can be reduced to defining the syntax of the trees, or equivalently the syntax of the symbolic expressions they represent. This is achieved by two sets, namely the functional and terminal sets. Elements of the latter are allowed as leaves of the tree representations, while elements of the first as internal nodes of the tree representations. An idea can be obtained by the logical example on the right-hand side of Figure 3.8, where the elements-values are at the leaves of the tree (terminals), while the operators (functionals) are internal nodes of the tree.

The chromosomes in genetic programming are characterized by two important features. First, they are non-linear structures, due to their tree nature. Secondly, their size is not fixed as in the approaches using linear structures, e.g., bit strings, but rather, the depth and breadth of the tree chromosomes in genetic programming representation may vary.

The main operators in genetic programming are recombination and mutation, even though the latter has received considerable skepticism specifically in genetic programming approaches. A characteristic difference compared to other approaches such as genetic algorithms and evolutionary strategies is that crossover and mutation are performed in one step, contrary to the other approaches where the crossover and mutation take place in two distinct steps. Crossover is applied on an individual by simply switching one of its nodes with another node from another individual in the population. With a tree-based representation, replacing a node means replacing the whole branch that the node belongs to, namely replacing the whole subtree rooted at the node replaced. This adds greater effectiveness to the crossover operator. The expressions resulting from crossover are very different from their initial parents. Mutation affects an individual in the population. It can replace a whole node in the selected individual, or it can replace just the node's information. To maintain integrity, operations must be fail-safe or the type of information the node holds must be taken into account. For example, mutation must be aware of binary operation nodes, or the operator must be able to handle missing values.

Mutation has two parameters, the probability p_m to choose mutation versus recombination and the probability to chose an internal point as the root of the subtree to be replaced. In general, it is advised that p_m is very small, almost zero. The most common implementation works by replacing the subtree starting at a randomly selected node by a randomly generated tree. Typically the newly created tree is generated the same way as in the initial population. This means that the size (tree depth) of the child can potentially exceed that of the parent.

Recombination in genetic programming creates offspring by swapping material among the selected parents. It is essentially a binary operator creating two child trees from two parent trees. A simple and most common implementation randomly selects two nodes of a tree parent and exchanges the subtrees of these nodes. Recombination in genetic programming has two parameters, the probability of choosing recombination or mutation, as in the mutation case, and the probability of choosing an internal point within each parent as crossover point.

Regarding parent selection, genetic programming usually employs fitness proportionate selection, or in cases of large population sizes, a method denoted by over-selection (typically for populations above 1000 individuals). In the latter the population is first ranked by fitness and then divided into two groups, one with the top $x\%$ and the other with the rest of the individuals. When parents are selected, 80% of the selection operations come from the first group and the other 20% from the second group. The value for the percentage of x is empirically found/tuned depending on the underlying population size.

For survival selection, genetic programming uses a strategy where the number of offspring created is the same as the population size and all the individuals have a life span of one generation (generational strategy with no elitism).

For the initialization of genetic programming the ramped half-and-half method is usually employed, where a maximum initial depth D_{max} of trees is specified and each member of the initial population is created from the available sets of functions and terminals according to two alternative methods, referred to as full and grow method, respectively. In the first, for each branch, which has depth D_{max}, the contents of nodes at depth d are chosen from the function set if $d < D_{max}$ and from the set of terminals if $d = D_{max}$. In the second, the branches may have different depths up to D_{max}. The tree is constructed starting with the root and the contents of each node are randomly chosen from the union of the set of functions and terminals if $d < D_{max}$.

An emerging effect is that the chromosomes tend to increase in size in the search process, a phenomenon referred to as bloat. The main reason for this is that the sizes are allowed to vary and within the survival of the fittest regime, chromosomes tend to increase in order to ensure they have a high value of the fitness function. A simple way to restrict the effect is to impose a maximum tree size and forbid a variation operator if the resulting children would exceed the defined maximum size. This threshold can be considered as one additional parameter of mutation and recombination, while another more sophisticated approach would impose a penalty term in the fitness formula for large chromosomes, thus driving them towards smaller sizes in order to maintain high fitness function.

A noteworthy aspect of genetic programming are the cases where executing a given expression of a parse tree representation changes the environment, which in turn affects the execution of the expression, and thus the fitness of the specific node. In such cases the solutions are not just simple mappings from the inputs to the outputs, which means that some additional information is required, perhaps in the form of a memory, to complete the mapping. This means that fitness evaluations become more demanding and they potentially require more time than in the simple case of data fitting problems, where the above mapping is straightforward. However, the quality of the evolved results is superior and this compensates for the higher evaluation cost paid.

3.4.4   Evolutionary Programming

Evolutionary programming is one of the four major evolutionary algorithm paradigms. It is similar to genetic programming, but the structure of the program to be optimized is fixed, while its numerical parameters are allowed to evolve. It was first used by Lawrence J. Fogel in the US in 1960 in order to use simulated evolution as a learning process aiming to generate artificial intelligence [66]. More specifically, intelligence was considered as the capability to adapt the behavior of the system, while meeting certain goals. In order to achieve this goal, prediction of the environment was considered essential, and in fact it was assumed to be a key feature for developing intelligence.

In the classical paradigm of evolutionary programming, Fogel used finite state machines as predictors. A Finite State Machine (FSM) is conceived as an abstract machine that can be in one of a finite number of states. The FSM is only in one state at any time and it has a finite number of possible state transitions from a state to another state. The FSM is also stimulated by a finite alphabet of input symbols and it can respond with a finite number of output symbols. An FSM can be used to implement learning and prediction functions. The fitness of an FSM can be defined as the prediction accuracy of the machine. Possible mutations of an FSM are changing an output symbol, changing a state transition (next state of the FSM), adding/deleting a state, and changing the initial state.

In this particular evolutionary approach, no recombination is employed, the representation involved real valued vectors, mutation is again a Gaussian perturbation, and parent selection is deterministic. Survivor selection is probabilistic and in meta-evolutionary programming self-adaptation of mutation step sizes is employed.

Currently evolutionary programming is a wide evolutionary computing variation with no fixed structure or representation, in contrast to some of the other variations of evolutionary computing. It is lately becoming harder to distinguish from the evolutionary strategies approach described in the previous section.

Evolutionary programming is most frequently used for optimizing multivariate real functions and in this case a straightforward floating-point representation is employed. Self-adaptation is used as a standard feature nowadays in evolutionary programming and thus a general representation usually employed is: $< x_1, ..., x_n, \sigma_1, ..., \sigma_n >$, where $\bar{x} = \{x_1, ..., x_n\}$ and $\bar{\sigma} = \{\sigma_1, ..., \sigma_n\}$ for the general form of individuals in evolutionary programming.

The main variation operator of evolutionary programming is mutation. In general, there is no single evolutionary programming operator, but rather the choice is determined by the application and especially the employed representation on a per case basis. Nevertheless, one of the most commonly used mutation operators, especially in the evolutionary variation, known as "meta-EP," uses self-adaptation of strategy parameters and a real-valued representation. Mutation in meta-EP transforms a chromosome $< x_1, ..., x_n, \sigma_1, ..., \sigma_n >$ into $< x'_1, ..., x'_n, \sigma'_1, ..., \sigma'_n >$, where $x'_i = x_i + \sigma' N_i(0, 1)$ and $\sigma'(1 + \alpha N(0, 1))$. $N(0, 1)$ denotes a random sample from a Gaussian distribution with zero mean and standard deviation 1, while $\alpha \cong 0.2$. A boundary rule to prevent standard deviation values close to zero is in general employed/suggested. Some of the differences for the self-adaptation of the step values that may be encountered in various evolutionary programming variations are:

- Varying the formula for the step sizes, e.g., using log-normal function as in evolutionary strategies;

- Incorporating variances instead of standard deviations as strategy parameters;

- The order in which the $\bar{\sigma}$ and the $\bar{x}$ are mutated.

Regarding the first and second, a common problem is the emergence of negative (invalid) values for the offspring variance, while the third regards the order of $\bar{\sigma}$, $\bar{x}$ updates, several studies of which have shown that it is more preferable to apply the "sigma first" strategy, which yields a more consistent general advantage over the "sigma last" strategy.

In general, recombination in evolutionary programming has withstood a long debate on the very essence of being useful for evolutionary programming, since individuals in the corresponding search space represent abstract species, rather than members of a single species, which in turn raises questions for the context of representation, namely what recombination of different species would mean. Until today, the question of whether crossover offers performance benefits for evolutionary programming remains open and it is a potential subject for further research in the future.

Regarding parent selection in evolutionary programming, this is much less an issue compared to other evolutionary algorithms. Here every member of the population creates exactly one offspring via mutation, a fundamental difference from genetic algorithms and genetic programming, where selective pressure based on fitness is applied. Evolutionary programming differs from evolutionary strategies, since the choice of parents in the first is deterministic, whereas in evolutionary strategies it is stochastic. Therefore, each parent generates an offspring (survivor selection), using a $(\mu + \mu)$ survivor selection. Sometimes stochastic variants are employed, where pairwise tournament competitions are held in the round-robin format and involve both the parent and offspring populations. The μ solutions with the greatest number of wins are retained to be parents of the next generation. This variant allows for less-fit solutions (individuals) to survive into the generation if they had a lucky draw of opponents. In the limit, the mechanism becomes deterministic as in the case of evolutionary strategies.

3.4.5   Evolutionary Computing at a Glance

Table 3.2 summarizes the features and special characteristics of the presented evolutionary algorithms described above. It may be compared with Table 3.1, in order to obtain a holistic overview of the common features, as well as differences between typical evolutionary algorithms.

A few emerging trends may be identified by the joint observation of these tables, mainly regarding the employed representations and mutations. The applicable problem sets are also closely related, verifying the fact that evolutionary approaches all share a common main approach that addresses broader problems emerging in various and diverse applications.

3.4.6   Parameter Control in Evolutionary Algorithms

For a full specification of the evolutionary computing algorithms presented before, a number of additional data requires clarification, in addition to the specific techniques employed for, e.g., representation, crossover, mutation, etc. Such data are called *algorithm parameters*. The values of these parameters greatly affect the behavior of the algorithms and eventually determine whether a specific algorithm will find an optimal or near-optimal solution, and whether this solution will be efficiently obtained. This subsection will summarize some of the typical approaches employed for the right selection of these parameter values.

In general, there are two major approaches for parameter setting, namely *parameter tuning* and *parameter control* . The first is a static approach where parameter values remain fixed for a running, but they can be set in successive algorithm runs according to a training approach, where parameters change in successive runs, until a convenient setting is obtained. This strategy is an experiment based approach, yielding suboptimal settings in most cases,

Table 3.2: Summary of evolutionary algorithm features.

Elements	Evolution Strategies	Evolutionary programming	Genetic programming
Typical problems	Continuous optimization	Optimization	Modeling
Representation	Real vectors	Real vectors	Tree structures
Recombination	Discrete/intermediary	None	Exchange of subtrees
Mutation	Gaussian perturbation	Gaussian perturbation	Random change in trees
Parent selection	Uniform random	Deterministic	Fitness proportional
Survivor selection	(μ,λ) or $(\mu+\lambda)$	Probabilistic $(\mu+\lambda)$	Generational replacement
Speciality	Self-adaptation	Self-adaptation	Machine learning

requiring substantial experimentation time, and hence, almost always it is deemed inappropriate for evolutionary algorithms, especially when the parameters are not independent of each other. In the second, parameter control, an initial set of parameters is selected and their values change during the execution of the algorithm. Namely, parameter control is an adaptive approach that dynamically alters parameter settings based on some learning, decision, or prediction approach, in order to achieve better overall performance and efficiency.

Several other approaches have been devised for proper parameter setting; however, most of them have been proven inefficient in practice, suffering from a series of problems (usually one approach solving one problem exhibits ill-behavior with respect to another). But this is exactly the type of problems that evolutionary algorithms were developed for. Consequently, it is natural to consider exploiting an evolutionary algorithm for tuning the parameter set of another evolutionary algorithm for a particular problem. The second evolutionary algorithm used for tuning the specific evolutionary algorithm under particular problem scenarios is called the meta-evolutionary algorithm. Of course the same evolutionary approach can be used for both tasks, namely self-tuning and problem solving. Self-adaptation used for varying mutation parameters is exactly representative of such an approach.

Changes in the parameters can be based on feedback from the search process, as in Rechenberg's 1/5 rule, which requires that the ratio of successful mutations to all mutations be 1/5, and hence, if the ratio is greater than 1/5 the mutation step size is increased, whereas if the ratio is less than 1/5 the step size is decreased. Apart from mutation, the fitness function itself (evaluation function) can be parameterized and varied less frequently. A very characteristic case of the latter is penalty function coefficient variation in constrained optimization problems, where static penalty weights (indicating the importance of violating a constraint) can become dynamic, depending on search feedback as in Rechenberg's 1/5 rule, or determined according to self-adaptation as in the mutation variation case. These three alternatives applicable to mutation and evaluation function are valid for any parameter of an evolutionary algorithm. Different domains of influence in the search space that the various parameters may have are called *scopes* (domains of influence).

In order to classify parameter control techniques, a number of factors has to be taken into account such as what is changed, how is it changed, the data based on which the change is carried out and the scope/level of change. Regarding the first, practically any component of an evolutionary algorithm can be parameterized. Representation, evaluation function, selector operators (parent and survivor), population, and mutation operators are some of the components that can be parameterized, as was shown before for mutation and the evaluation function (penalty weight).

Regarding the "how" changes are made question, three main approaches have been mentioned already, namely deterministic parameter control, adaptive parameter control, and self-adaptive parameter control. In *deterministic parameter control* the value of a strategy parameter is altered by some deterministic rule that modifies parameters in a fixed predetermined manner. *Adaptive parameter control* is when some form of feedback from the search that serves as inputs to a mechanism used to determine the direction or magnitude of the change to the strategy parameter is implemented. Finally, in *self-adaptive parameter control* the parameters to be adapted are encoded into the chromosome and undergo mutation and recombination, essentially implementing an evolution of the evolution concept. Regarding the deterministic characterization of the first approach on "how" to implement changes, it should be noted that it refers to the fact that the parameter-altering transformations take no input variables related to the search process. In that sense, the term "fixed" parameter control might better reflect the actual mechanics behind this class of approaches.

In self-adaptive parameter control, determining changes on the parameter set is based on a monitoring system that gathers information on the evolution/behavior of the parameters and the evolutionary algorithm progress and it is used as feedback for adjusting the parameters. Two approaches have emerged, the first with absolute evidence and the second with relative evidence. *Absolute evidence* characterizes the cases where the value of a strategy parameter is altered by some rule that is applied when a predefined event occurs. However, this trigger is not deterministic as in the deterministic parameter control, but rather it is based on feedback from the search process. Examples include increasing the mutation rate when population diversity drops below a threshold, or changing the probability of applying mutation/crossover according to a fuzzy rule, and it is clear that such mechanisms require that the user has a clear (empirically or theoretically obtained) intuition about how to steer the given parameter into a certain direction in cases that can be specified in advance. In *relative evidence* parameter values are compared according to the fitness of the offspring that they produce and the better values get rewarded. In this case, the direction and/or magnitude of the change of the strategy parameter is not deterministically specified, but is specified relative to the performance of other values.

The scope/level of a change determines whether any change within any component of an evolutionary algorithm will affect other subcomponents and the corresponding degree. Naturally, this depends on the component of the evolutionary algorithm and the underlying degrees of freedom it allows. In that sense this is a secondary feature of parameter control, which is component and implementation dependent.

Summarizing the above, the main classification criteria for parameter setting in evolutionary algorithms during the execution of the algorithm (parameter control) are:

- What component/parameter is changed (representation of individuals, evaluation function, variation operators, selection operator, replacement operator and population)?

- How is the change made (deterministic, adaptive or self-adaptive)?

Table 3.3: Summary of evolutionary algorithm features.

	Deterministic	Adaptive	Self-adaptive
Absolute	+	+	
Relative		+	+

- Which evidence is employed for making the change (absolute or relative evidence)?

Thus, a complete classification of parameter control is three dimensional, where the dimensional component has six alternatives, and the other two dimensions have three and two alternatives, respectively. The available combinations for the last two dimensions (on the change-evidence plane) are provided in Table 3.3. Figure 3.9 depicts cumulatively the overall classification of parameter setting approaches.

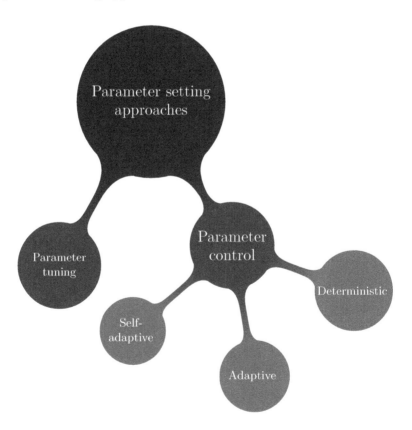

Figure 3.9: Taxonomy of parameter control approaches. Parameter tuning takes place before execution, while parameter control takes place during the execution.

3.4.7   Special Forms of Evolution

Apart from the evolution approaches described up to this point in this chapter, all of which follow an overall methodology described in Subsection 3.3.1, there are some other forms of evolution that deviate from the standard evolutionary algorithm methodology. In particular, co-evolution and interactive evolution are two characteristic representatives of these alternative evolutionary approaches, both of which operate under the "external influence concept," where a population or a person are employed respectively to affect the fitness of the main population, and thus its evolution. In both cases, the fitness value awarded to a solution may vary due to the fitness function being dependent on the evolution of another population in the co-evolution case, or due to inconsistencies exhibited by the subjective preferences of the fitness-affecting users.

Co-evolution

Co-evolution was inspired by the fundamental observation that in real life, the fitness of an organism is entirely determined by the environment in which it resides. The characteristics of this environment are predominantly determined by the presence and behavior of other organisms from the same or different species (competition). Thus, the effect of other species in determining the fitness of an organism can be positive or negative. The terms *mutualism* and *symbiosis* are employed to denote co-adaptation of species in a mutually beneficial way and the terms *predation* or *parasitism* refer to cases of one species having negative effects on the evolution of the other. However, the main notion is that since evolution affects all species, the net effect is that the perceived landscape for each species is affected by the configuration of all the other interacting species. This is referred to as co-evolution.

There are two main variants of co-evolution computational models, cooperative and competitive co-evolution. Each of these variants is discussed separately in the following.

Co-evolutionary models where a number of different species, each representing part of the problem, cooperate in order to come up with a solution to a larger problem are referred to as cooperative co-evolution. The main advantage is that it allows for effective decomposition of problems. Essentially, each subpopulation is solving a (much) smaller and more tractable problem. If the smaller problem is not tractable, then there is no point in doing such decomposition and an alternative approach should be attained. However, on the other hand, a suitable partition must be provided and usually in an automated manner.

In an extreme point of cooperative co-evolution, two species become so interdependent on each other that they end up linked to each other, a situation referred to as endosymbiosis. The extent to which such populations, which in evolutionary computing terms represent two different types of solutions, affect the emerging strategies, depends significantly on the degree of such linkage, so that eventually solutions with high linkage between the two populations are preferred.

When cooperating co-evolution is employed, one of the significant aspects is the way that solutions obtained from one population are combined with solutions obtained from the other populations in order to obtain a meaningful fitness evaluation. Various alternatives proposed in the literature include using generational genetic algorithms in each subpopulation, using lifetime fitness evaluation in the steady state generational model, or using diffusion models of evolutionary algorithms, while changing the rate at which the composition of different populations is perceived by the user to change. Another employed approach is the use of automatically defined functions within genetic programming, where the function set is extended to include calls to functions that are themselves being evolved in parallel and separate populations. All these approaches introduce additional diversity and add in the "stochastic search" degree of the developed approach in each case.

On the other hand, in competitive co-evolution, individuals compete against each other to gain fitness at each other's expense. Again the individuals may belong to the same or different species. As with cooperative co-evolution, the fitness values for the various populations will change along with their evolution and once more, the selection of pairing strategies among solutions from individual populations can have a significant impact on the overall outcome. There are two cases emerging, one where competition arises within a single population (species) and then each strategy can be paired against the other, or a randomly chosen sample of the others, and a second one where competition arises between different populations and then a pairing strategy is necessarily required for evaluating fitness.

The driving force of co-evolution is usually denoted as *competitive fitness evaluation* [9]. The most prominent feature of competitive fitness evaluation is that it is self-scaling, namely early in the operation of the process relatively poor solutions may survive, since their competitors have not grown strong yet. However, as the evolution progresses and the average strength of the population increases, the difficulty of the fitness function continually scales.

Interactive Evolution

The characteristic feature of interactive evolution is that the user applying the approach, i.e., the user, becomes part of the system playing the role of an oracle that guides the evolutionary process. The net result is that the yielded outcomes will better suit the expectations of the user and thus they will be considered more successful.

The two basic mechanisms of interactive evolution are *variation* and *selection*. Interactive evolution is especially concerned with the second. In general, the user can influence/control selection in various ways. The influence can be direct or indirect. In all cases, the user's influence is named subjective selection. The advantages of incorporating human guidance in the evolution are:

- Handling situations with no clear fitness function, e.g., in cases where the preference for certain solutions cannot be formalized.

- Improved search ability, where the user directs the search in case it traps in a local optimum.

- Increased exploration and diversity.

However, these advantages come at a cost too, namely:

- Slowness, compared to fully automated processes.

- Inconsistency, since users may change their minds on the fly.

- Limited coverage, since the users cannot handle large populations effectively.

Interactive evolution is usually related to evolutionary design and component-based representations. It is oriented towards exploring the search space. The basic template for an interactive evolutionary design system consists of five components:

1. A phenotype representation defining the application-specific kind of objects that are to be evolved.

2. A genotype representation, where the genes represent (directly or indirectly) a variable number of components making up a phenotype.

3. A decoder that defines the mapping from genotypes to phenotypes.

4. A solution evaluation facility allowing the user to perform.

5. An evolutionary algorithm to perform the actual search.

Such schemes can be used to evolve objects in a great variety that cannot be obtained by typical approaches. These systems resemble their evolutionary counterparts a great deal. Their main difference lies in the fact that their objective is to please the user, rather than achieve some specific objective.

Chapter 4

Complex and Social Network Analysis: Metrics and Features

Exploiting features from Social Network Analysis (SNA) in Complex Network Analysis (CNA) and control, as explained in previous chapters, requires quantifying such features in accurate manners and then evaluating the effectiveness of the approaches with proper and measurable means. As mentioned already, these elements from network science could be used for improving several aspects of network performance and in various application settings, i.e., in different applications and diverse network structures. Such elements from SNA and network science should be measured and assessed in a meaningful, scalable and computationally efficient manner in order to be successfully identified first and then employed in the various application frameworks. Defining and exploiting the appropriate evaluation metrics that will be applied in the corresponding analysis and development mechanisms is also an important part of social/complex network analysis. In fact it is one of the first steps a designer should take in order to determine appropriately the effectiveness and efficiency of the developed mechanisms.

In this chapter, we will present and analyze the most prominent evaluation and assessment metrics emerging in complex/social network analysis, most of which will also be exploited in later chapters of this book, where network topology control and improvement mechanisms will be presented, focusing on communications networks. Several of the salient features of the analyzed metrics will be provided, thus serving as guidelines for further exploiting complex network analysis concepts in communications networks.

The included metrics have been inspired in many cases by social incentives and studies. These metrics are highly characteristic of social features/properties and could be used to assess the degree to which social mechanics/dynamics have been incorporated into communication networks, as will be shown in detail in later chapters. In addition, these metrics will be the means both for designing mechanisms that take advantage of these emerging social aspects of communications networks in order to improve performance, and the assessment of the successfulness of these mechanisms in achieving their goals. These paradigms will serve as starting points for developing similar frameworks and mechanisms that exploit other types of features from network science, e.g., elements from biological networks.

The majority of the complex network analysis metrics that will be presented in this chapter are centered around the notion of distance over metric spaces. Most frequently, the considered metric space is that of the discrete graph space, in which case, distances are measured in hops. Different metrics (or space embedding) have been proposed as well and will also be examined in this chapter. Additional details on computing distances in metric spaces are provided in Appendix B, in conjunction with the notion of semirings and path computations.

4.1   Degree Distribution

The notion of node degree has been explained in earlier chapters. It represents neighboring relations between a node and those interacting directly with it. The distribution of node degrees, commonly referred to as degree distribution, describes cumulatively such measures of direct neighboring relations for network nodes. Occasionally, the node degree depends on the Euclidean distance of nodes (if they are embedded in a Euclidean plane), i.e., in geometric and random geometric graphs [128]. In these cases, the degree distribution is an implicit derivative of a distance-based metric in the Euclidean space. However, in the general case, neighboring relations are determined according to different criteria, e.g., online social networks, protein networks, etc., and thus the degree distribution is derived from non-distance-based metrics.

The degree distribution could have a deterministic or probabilistic form, depending on the network analysis framework and the application scope of the studied/analyzed networks. In cases where neighborhood relationships do not vary, e.g., a snapshot of a local area network, or an edition of a transportation map, the degree distribution is the full spectrum of node degree values. It can be visualized by providing the value of each node degree for all nodes, as shown for different types of networks in Figure 4.1. In cases where network connections vary with time or other parameters and node connectivity cannot be deterministically defined, i.e., where only a connection probability between a node pair can be defined, the degree distribution becomes stochastic as well. In these cases, the degree distribution $P(k)$ is the probability that a node has k neighbors.

The above definitions properly describe networks with undirected topologies. However, the definitions need to be extended in order to cover the cases

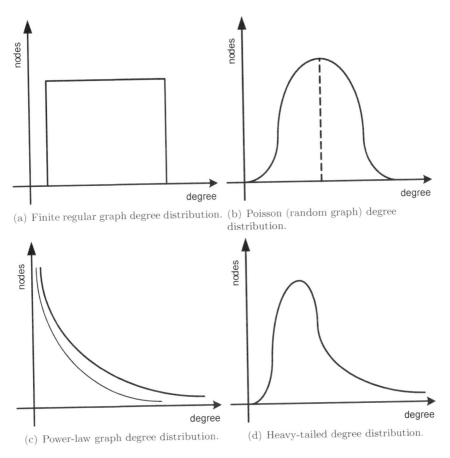

(a) Finite regular graph degree distribution. (b) Poisson (random graph) degree distribution.

(c) Power-law graph degree distribution.   (d) Heavy-tailed degree distribution.

Figure 4.1: Examples of degree distributions for various types of networks (linear scale).

of directional topologies as well. In undirected networks a single degree distribution suffices to characterize the whole network. In directed networks, two different distributions are required, one for the in-degree and one for the out-degree of the nodes. Each of the two distribution types is similar to the one for undirected graphs (could be deterministic or probabilistic, etc.) and provides similar information for the two asymmetric directions of flow, towards and from network nodes.

At this point, we need to note that in some works, the degree distribution is incorrectly referred to as connectivity of the network. This is a practice that should be avoided in general, since connectivity describes cumulatively the interconnection relations/capabilities of a network, i.e., in the sense of $\kappa(G)$, if G is the network graph (Section 2.1.3). The term distribution describes in a cumulative fashion properties of nodes individually, and thus the degree distribution should not be mistakenly used to characterize the global connectivity properties of a network.

The degree distribution is characteristic of a network topology. It essentially identifies uniquely a specific set of node interconnections and thus it segregates among different network types. In complex network analysis, the degree distribution is widely used for network classification. In fact, all network types that will be presented in the following chapter are segregated according to their interconnection relations, which in turn were reflected in their node degree distributions.

Especially in social networks, degree distributions are mainly characterized by heavy tails, i.e., in a chart of node degree-number of nodes (as the ones shown in Figure 4.1), the distribution is similar to that of a power-law in linear scale, where the curve exhibits a long tail in the axis of node degree, thus signifying the fact that a large number of nodes have small degree. This does not necessarily imply a typical power-law distribution, but rather that significant mass can be found towards that part of the distribution. Various heavy-tail distribution models maybe utilized in such cases for accurately modeling such types of networks.

However, obtaining accurate representations of degree distributions is not always feasible or at least efficiently viable, especially for real networks where observations may be fuzzy and inaccurate. Numerical methods and sampling are usually employed for reconstructing the degree distributions of real networks, in which case such distributions are usually referred to as *empirical degree distributions*. In fact, obtaining an accurate mathematical model for the degree distribution could be one of the toughest tasks, and thus, in many cases, the empirically obtained degree distribution might not offer significant benefits for the analysis of a network, but rather only a first indication of the interconnection nature and node behaviors. Other more efficient and indicative metrics should be considered, as shown in the following parts of this chapter.

At this point we need to note that the degree distribution is sometimes referred to as degree sequence. The latter is a characterization more suitable for deterministic graphs, where the node degree does not vary for each node, and thus the spectrum of node degrees may be indeed considered a sequence of possible degree values. The term degree sequence is not so appropriate for random or randomized graphs, where the spectrum of node degrees indeed comes at a specific probability distribution. In this book, we employ the term degree distribution for all cases, in order to have a common and uniform proper terminology. The main use of the degree sequence representation is reserved, as in many other works, for denoting an ordering relation, in the sense that the degree sequence is ordered in a monotone (increasing or decreasing) fashion, which in turn enables better analysis or description of the degree spectrum.

4.2   Strength

As already explained in previous chapters, weighted graphs emerge oftentimes in social and communications networks, especially in the latter, to signify intensities of the developed relations represented by the network edges. However, these intensities are not conveyed by the degree distribution described before, which only provides connectivity information. A more involved metric is required to convey this kind of information, which will be described in this and in the following sections.

The above concern becomes more prominent in weighted network representations, where the weights of the links make notable difference in various networking functions and mechanisms, e.g., shortest path computation and routing in communications networks. The weights assigned to connecting links (or sometimes nodes) also bear characteristic information for quantities exchanged between network nodes (e.g., traffic, information, money, shipments, proteins, signals, etc.). Such information should be taken into account in a more elaborated metric that provides the combined information of connectivity and intensity of connectivity for both undirected and directed networks.

Assuming a weighted undirected network graph, in addition to the adjacency matrix defined in Section 2.1.1, a weight matrix is employed as well. The weight matrix $W = [w_{ij}]$ is an $N \times N$ matrix, in which element w_{ij} specifies the weight in the link connecting vertices i and j ($w_{ij} = 0$ if the two vertices are not connected). Along with the degree of a node, a very useful and important measure of network properties in terms of node interaction is the *vertex strength* s_i:

$$s_i = \sum_{j \in \mathcal{V}(i)} w_{ij} \tag{4.1}$$

where the sum spans all neighbors of node i, in the neighborhood set $\mathcal{V}(i)$.

For a directed graph two strength measures need to be quantified, in order to relate the directionality information together with connectivity information and intensity of connectivity. Consequently, in-strength and out-strength measures are defined:

$$s_i^{\text{out}} = \sum_{j \in \mathcal{V}(i)} w_{ij} \tag{4.2}$$

$$s_i^{\text{in}} = \sum_{j \in \mathcal{V}(i)} w_{ji} \tag{4.3}$$

Eventually, node strength integrates in a uniform fashion connectivity and link weight information into a generalized connectivity metric that takes into account the "amount" of exchanged information between interacting agents-nodes. By combining vertex strengths, node strength can be turned into a network metric, indicative of the components of a network and possible partitions of nodes [155].

4.3   Average Path Length

Another very useful metric for performance evaluation in traditional communications networks and social structure assessment is the average path length. The average path length is a network-wide defined metric (compared to node degree, which is a distributed metric characterizing a specific node), which is defined as the average of the shortest path lengths between all pairs of nodes in the network. Thus, in order to obtain the average path length of a network, one needs to identify all possible node pairs (at most $\binom{|V|}{2}$) in the network, then compute the length of the shortest path for each node pair and average over all such pairs. By definition, the average path length requires knowledge of all of the shortest paths of a network, and thus average path length is an inherently centralized operation. However, in the literature several distributed approaches have been proposed for approximating the average path length within acceptable bounds of the accurate value and in meaningful time (Appendix B, [16], [112], [51], [26], [48] etc.).

The average path length indicates the expected distance one would experience between a randomly selected node pair of a network, according to the selected distance measure for computing shortest paths. Thus, it can be indicative of several performance aspects of the network for other metrics and operations that depend on the distance between nodes. However, as it can be realized from the preceding discussion, a very fundamental and critical aspect for computing the average path length of a graph is the definition of distance itself employed for the computation of the shortest paths. The definition of the distance among nodes, based on which neighboring relations between nodes are also usually determined, depends on the specific nature, purpose, operation, and application framework of a network. The most frequent metric employed is the hop-count, denoting the least possible number of links separating two nodes. Alternative metrics may include sum/product of the weights of the links along the path, if weighted networks are involved, or even other functions of link weights.

Especially for network graphs, the distance has usually been incorporated in the definition of the links between nodes. Thus, the actual distance between nodes is in turn determined by the hop-distance metric, indicating that according to the specific underlying distance, the nodes have an immediate (1-hop) distance, second order (2-hop) relations, etc.

The previous discussion essentially highlights the importance of the underlying employed algebra for computing node distances, imposed either by the operation and nature of a network or the design objectives of engineers. For instance, depending on the application, e.g., trust computation or traffic management, the shortest path may be more appropriate to be computed as product or sum of the link weights, thus yielding different average path length values for the same network topology.

In any case, the average path length is indicative of the relative spatial and connectivity potential spread across a network, and it is often used in the definition of more complex network evaluation metrics, such as centrality, as will be explained later. This confirms it as a very important assessment metric and one that network scientists should always take into account in their analysis.

4.4   Clustering Coefficient

The clustering coefficient is an important metric for complex networks, used to characterize the structure of a social network both locally, i.e., at the node level, and globally, i.e., at the network level. It computes the cliquishness of the network, expressing the degree of the triadic closure process in the corresponding network. The triadic closure [57] process takes place when two neighbors of a node become themselves neighbors and is very likely to happen in social networks. More explicitly, suppose having a friendship network, i.e., nodes are persons and links are formed between two nodes if they know each other on a first name basis. Then it is more likely that a new edge is formed between a pair of nodes (i, j) having a common neighbor k as k can bring in contact nodes i, j either willingly or randomly. The triadic closure is a process characterizing the network evolution; however, the clustering coefficient is computed on static networks. As a result, by computing the clustering coefficient at a particular time instant, we can get a feeling of how much triadic closure has happened till then in the network, i.e., the higher the clustering coefficient, the more the participation of the triadic closure in the network evolution. It should be mentioned that the clustering coefficient of a random graph $\mathcal{G}(n, p)$, defined in Section 2.2.1, equals p, since the probability that two neighbors of a node are directly connected (and thus a triangle is formed) equals the probability p that an edge exists in the graph. In the sequel, we quote two definitions of the clustering coefficient in binary, undirected graphs; the first one defines a local clustering coefficient for each node and then generalizes it to the whole network, while the second definition defines a global clustering coefficient for the whole network. Afterwards, we extend the notion of the clustering coefficient in weighted and directed networks.

4.4.1   Definition

The local clustering coefficient C_i of node i is a measure of direct connectivity between the neighbors of node i:

$$C_i = \frac{\text{number of triangles connected to node } i}{\text{number of triples centered at node } i} \quad (4.4)$$

$$= \frac{\text{number of edges between the neighbors of node } i}{\text{number of all possible edges between the neighbors of node } i} \quad (4.5)$$

In graph theory, a triangle is an ordered, complete 3-node subgraph, while a triplet is simply a connected 3-node subgraph. The network local clustering coefficient is the average $C_{net} = \frac{1}{N} \sum_{i=1}^{N} C_i$ over all network nodes N. The local clustering coefficient at node i has a value field equal to $[0, 1]$, where $C_i = 0$ when there are no connections among the neighbors of i (zero number of triangles connected to i) and $C_i = 1$ when there exist all the possible connections among node i's neighbors. If i has k_i neighbors, then the number of all possible connections among node i's neighbors is equal to $\frac{k_i(k_i-1)}{2}$.

If we consider that $A = [a_{ij}]$ is the $N \times N$ adjacency matrix of the network, as this is described in Section 2.1.1, Eq. (4.5), through simple matrix equations, can take the form:

$$C_i = \frac{\sum_{j,k} a_{ij} a_{jk} a_{ki}}{(\sum_j a_{ij})^2 - \sum_j a_{ij}} = \frac{(A^3)_{ii}}{(\sum_j a_{ij})^2 - \sum_j a_{ij}} \tag{4.6}$$

i.e., the number of triangles adjacent to node i is equal to $T_i = \frac{\sum_{j,k} a_{ij} a_{jk} a_{ki}}{2}$. The global clustering coefficient C_G of a network measures the portion of triangles versus triples in the whole network topology:

$$C_G = \frac{\text{number of triangles in the network}}{\text{number of triples in the network}} \tag{4.7}$$

To illustrate the above definitions, Figure 4.2(a) depicts a simple network topology for which we are going to compute the local and global clustering coefficients. Figure 4.2(b) shows an indicative triangle (g-h-j) and an indicative triplet (i-k-l). Firstly we compute the local clustering coefficients for each one node separately. Node i has 3 neighbors, the maximum possible number of connections between them is $\frac{3*2}{2} = 3$, and there exists only 1 of them, i.e., $C_i = \frac{1}{3}$. Node k has two neighbors that are not connected, therefore, $C_k = 0$. Similarly, it can be verified that $C_j = \frac{1}{3}$, $C_l = 0$, $C_h = \frac{2}{3}$, $C_g = 1$. The average local clustering coefficient over the whole network is $C_{net} = \frac{C_k + C_i + C_j + C_l + C_h + C_g}{6} = \frac{0 + \frac{1}{3} + \frac{1}{3} + 0 + \frac{2}{3} + 1}{6} = \frac{7}{18}$. Regarding the global clustering coefficient, the network as a graph has 15 connected triples and 6 triangles, i.e., $C_G = \frac{6}{15}$.

4.4.2   Extension to Weighted Graphs

As it is commonly accepted, a binary graph representation is not sufficient for capturing the complex features of modern interactions. Weighted network graphs can replace binary graphs for including information about the intensity or importance of the communication between two nodes. As a result each link (i, j) is associated with one or more weight values expressing the corresponding performance metrics of interest such as cost of communication on the link (i, j), trust value that node i assigns to node j, etc. In this section we extend the definition of the local clustering coefficient to weighted graphs, based on

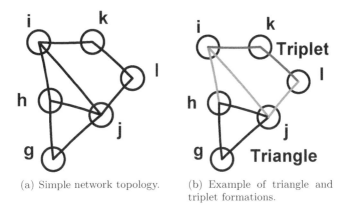

(a) Simple network topology.

(b) Example of triangle and triplet formations.

Figure 4.2: Example of the computation of the clustering coefficient.

proposed extensions from related work. As described in Section 4.2, let us assume that the network graph is represented by the weight matrix $W = [w_{ij}]$ with $w_{ij} = 0$ if the edge (i, j) does not exist. The main idea of this extension is not only taking into account the number of triangles and triples connected to each node, but also consider the importance of each one triangle or triplet as this is defined based on the weight values of the links they consist of. There are many definitions of the clustering coefficient for weighted networks in the related bibliography [17, 88, 118, 120, 143], where each one computes a different function of the links' weights included in each triplet or triangle leading in this way to a different measure of importance of the triangles/triplets. Here we quote and explain three representative of them [143].

To begin with, Barrat et al. [17, 143] defined the "weighted local clustering coefficient" as:

$$Cb_i^w = \frac{\sum_{j,h} \frac{(w_{ij}+w_{ih})}{2} a_{ij} a_{ih} a_{jh}}{s_i(k_i - 1)} \qquad (4.8)$$

According to Eq. (4.8), if $a_{ij} a_{ih} a_{jh} = 1$, i.e., the edges (i, j), (i, h), (j, h) exist, the sum $(w_{ij} + w_{ih})/2$ of the weights of the links of the triangle among i, j, h, adjacent to node i is added to the nominator of Cb_i^w. As a result, for the computation of the weighted clustering coefficient Cb_i^w, it takes into account for each triangle among i, j, h, the intensity of interaction of the triplet among j, i, h centered at i. The denominator serves the purpose of normalization so that $0 \leq Cb_i^w \leq 1$, which is more clear if it is written in the form of $s_i(k_i - 1) = \bar{w}_i k_i(k_i - 1)$, where $\bar{w}_i$ represents the average weight of the edges adjacent to node i, since $s_i = \sum_j w_{ij} = \bar{w}_i k_i$. If all the weights (on existing edges) are equal to 1 (case of binary graph), or if all the weights are equal to a constant value, Eq. (4.8) reduces to the Eq. (4.6). If $Cb_i^w = 0$, there are no connections among the neighbors of node i. Also, $Cb_i^w = 1$ if the binary clustering coefficient $C_i = 1$, i.e., if there exist all the possible links between the neighbors of node i.

Another definition of the local clustering coefficient in weighted networks is proposed in [143, 169], and constitutes a generalization of Eq. (4.6) by replacing the adjacency matrix A with the weight matrix W and changing the denominator so as to become an upper bound for the nominator ensuring that the local clustering coefficient is less than unity. Also, in order to apply Eq. (4.9), we need to normalize the weight values by dividing each weight w_{ij} with the maximum weight, i.e., $\max_{i,j} w_{ij}$, where we denote the fraction as $\hat{w} = \frac{w_{ij}}{\max_{i,j} w_{ij}}$. Therefore,

$$Cz_i^w = \frac{\sum_{j,k} \hat{w}_{ij} \hat{w}_{jk} \hat{w}_{ki}}{(\sum_j \hat{w}_{ij})^2 - \sum_j \hat{w}_{ij}^2} \tag{4.9}$$

In the case of binary graph, Eq. (4.9), reduces to Eq. (4.6) since $W = A$ and $\sum_j \hat{w}_{ij}^2 = \sum_j a_{ij}^2 = \sum_j a_{ij}$. In addition, $Cz_i^w = 0$ when there are no connections among the neighbors of node i. The value $Cz_i^w = 1$ is achieved when there exists all the possible connections among the neighbors of node i (i.e., when $C_i = 1$) and for each triangle among i, j, k it holds that $w_{jk} = \max_{i,j} w_{ij}$.

Last but not least, we quote a definition for the clustering coefficient in weighted graphs, which regards that each formed triangle, including node i, contributes to the nominator of the weighted local clustering coefficient of node i a quantity equal to the geometric mean of the weights of the edges of this triangle [118, 143]. Again, the weights should be normalized with respect to their maximum value as in the previous case and they are denoted again as $\hat{w}$. The corresponding equation is:

$$Co_i^w = \frac{\sum_{j,k} (\hat{w}_{ij} \hat{w}_{jk} \hat{w}_{ki})^{\frac{1}{3}}}{k_i(k_i - 1)} \tag{4.10}$$

Similarly to the previous cases, $Co_i^w = C_i$ in the case of a binary graph and $Co_i^w = 0$ in a case where the subgraph of the neighbors of node i consists of k_i disconnected components. Finally, $Co_i^w = 1$, if the subgraph of the neighbors of i is complete and all the edges participating at triangles have maximum weights, i.e., equal to $\max_{i,j} w_{ij}$.

4.4.3 Extension to Directed Graphs

Many real world complex networks include connections deriving from non-mutual relationships. In graph theory language this is expressed as follows: if the link (i, j) exists in the graph, the link (j, i) may exist or not (as also described in Chapter 2). As an example, in the online social network Twitter, if node i is a follower of another node j this does not mean that j will reciprocate this relationship and become a follower of i. As a result, complex networks are modeled in higher detail if directed (and weighted) graphs are used. In this section we are going to study the extensions of the clustering coefficient for directed binary and weighted graphs as these are described in [63]. One possible way of treating the clustering coefficient measure in directed network graphs would be to symmetrize the network and use one of the definitions described in the previous sections. This approach, although easy to implement, leads to the loss of important information regarding the information flow in the network, the structure of the network links, and their formation. Based on this fact we focus on the non-trivial extensions of the definition of the clustering coefficient in directed network graphs, which take into consideration the direction of the edges in the definitions of the triangles and the triplets.

Before quoting the definitions, we define some notions that are going to be used [63]. The in-degree and the out-degree of a node i are defined correspondingly as $k_i^{in} = \sum_{j=1}^{N} a_{ji}$, $k_i^{out} = \sum_{j=1}^{N} a_{ij}$ (Chapter 2). Moreover, the total degree of node i is defined as $k_i^{tot} = k_i^{in} + k_i^{out} = \sum_{j=1}^{N} (a_{ij} + a_{ji}) = (A^T + A)\mathbf{1}$, where A^T is the reverse matrix of A and $\mathbf{1} = [1, 1, 1, ..., 1]^T$, a N-dimensional column vector. Let us also consider that the number of links adjacent to i existing in both directions is denoted as k_i^b, i.e., $k_i^b = \sum_{j=1, j \neq i}^{N} (a_{ij} a_{ji}) = A_{ii}^2$. If we symmetrize the network graph by considering each one of its connections as bidirectional, the induced node degree of i will become equal to $k_i = k_i^{tot} - k_i^b$, since each one of its bidirectional connections will be counted twice. Similar definitions hold in the case of weighted networks where instead of using the node degree we use the node strength, as this is defined in Section 4.2.

In a directed network, node i can form 8 different triangles with any pair of its neighbors (2^3, since each one of the three edges of the triangle has two possible directions). Let us use the term "directed triangle" to refer to one of these. All the possible directed triangles centered at node i are visualized in Figure 4.3. The more intuitive extension of the definition of the clustering coefficient to directed graphs is the fraction of the directed triangles actually formed among i and its neighbors (t_i^D), towards all the possible directed triangles that can be formed among i and its neighbors (T_i^D), i.e.

$$C_i^d = \frac{t_i^D}{T_i^D} \tag{4.11}$$

$$= \frac{\sum_{j,h:j \neq h \neq i} (a_{ij} + a_{ji})(a_{ih} + a_{hi})(a_{hj} + a_{jh})}{2(k_i^{tot}(k_i^{tot} - 1) - 2k_i^b)} \tag{4.12}$$

$$= \frac{(A + A^T)_{ii}^3}{2(k_i^{tot}(k_i^{tot} - 1) - 2k_i^b)} \tag{4.13}$$

where in Eq. (4.12), the nominator counts all directed triangles, which becomes more evident if it gets expanded to a sum of products, each one representing a different triangle (i.e., products of the form $a_{ij} a_{ih} a_{jh}$). The denominator

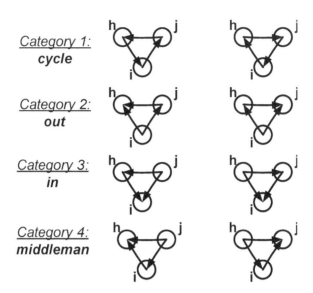

Figure 4.3: All possible directed triangles formed among i and two of its neighbors j, h.

expresses all possible triangles formed among i and its neighbors due to the fact that node i can form up to $2\frac{k_i^{tot}(k_i^{tot}-1)}{2}$ directed triplets with its neighbors and each triplet corresponds to two possible directed triangles (depending on the direction of the third edge). However, $2k_i^b$ among them are formed between i and the same neighbor j, in the cases that both edges (i,j), (j,i) exist, and therefore, they are wrongly considered and should be subtracted. Moreover, C_i^d reduces to C_i (Eq. 4.6) if the graph A (adjacency matrix) is binary, as can be easily proved by replacing the reciprocated connections $(a_{ij} = a_{ji} \; \forall \; i,j)$. Finally, the directed clustering coefficient of the whole network is defined as the average of the C_i^d for all i as exactly defined for binary undirected graphs, i.e., $C_{net}^d = \frac{\sum_i C_i^d}{N}$. For weighted and directed networks the Eq. (4.13) is extended by replacing the adjacency matrix A with the weight matrix $W^{\frac{1}{3}} = \left[w_{ij}^{\frac{1}{3}}\right]$ as follows:

$$C_i^{wd} = \frac{(W^{\frac{1}{3}} + W^{\frac{1}{3}T})_{ii}^3}{2(k_i^{tot}(k_i^{tot}-1) - 2k_i^b)} \qquad (4.14)$$

If the graph W is symmetric, i.e., all the consisting edges are bidirectional with equal weights on both directions, Eq. (4.14) reduces to Eq. (4.9). Furthermore, if the graph is directed and binary, Eq. (4.14) becomes equal to Eq. (4.13).

In the sequel, more specialized definitions of the clustering coefficient are going to be presented. The definition of Eq. (4.13) counts all triangles regardless of the directions of their edges. However, the direction of the triangle edges matters as it determines the flow of information in the network and by observing Figure 4.3, all the possible directed triangles can be divided in 4 patterns or categories. The category "cycle" contains the triangles consisting of edges with directions that form a directed circle of length 3. The category "in" contains the triangles two edges of which point at node i. Similarly, the category "out" consists of triangles two links of which leave node i. Finally, for the category "middleman" the nodes j, h are connected directly or through a directed path of length 2 via node i. For each one of these categories, a separate clustering coefficient can be defined [63]. Let us denote the number of triangles of the category p, $p = 1, 2, 3, 4$ actually formed by i and its neighbors as t_i^p, and the maximum possible number of triangles of the category p that can be formed by i and its neighbors as T_i^p. As a result the local clustering coefficient at node i and for the p category is defined as

$$C_i^p = \frac{t_i^p}{T_i^p} \tag{4.15}$$

As a result for computing C_i^p, $\forall\, p$, we need to compute t_i^p, $T_i^p\ \forall\, p$. We begin by computing $T_i^p\ \forall\, p$ by analyzing the equation for the maximum possible number of directed triangles that can be formed by node i (denominator of Eq. (4.13)) as follows:

$$
\begin{aligned}
k_i^{tot}(k_i^{tot} - 1) - 2k_i^b &= (k_i^{in} + k_i^{out})(k_i^{in} + k_i^{out} - 1) - 2k_i^b \\
&= (k_i^{in} k_i^{out} - k_i^b) + (k_i^{out}(k_i^{out} - 1)) \\
&\quad + (k_i^{in}(k_i^{in} - 1)) + (k_i^{in} k_i^{out} - k_i^b) \\
&= T_i^1 + T_i^2 + T_i^3 + T_i^4
\end{aligned}
\tag{4.16}
$$

In the case of T_i^1, T_i^4, k_i^b should be subtracted for the same reason explained earlier, i.e., the product $k_i^{in} k_i^{out}$ counts triangles formed by i and two opposite directed edges with the same neighbor. Also, T_i^1, T_i^4 have the same form, as they differ only by the orientation of the third edge (i.e., between nodes j, h in Figure 4.3), which does not affect the maximum number of triangles for each category. In the sequel, $t_i^p\ \forall\, p$ are computed through simple computations using the elements of the adjacency matrix A [63]. Therefore, for the triangles belonging to the category cycle, we have:

$$t_i^1 = \frac{1}{2}\sum_{j,h}[a_{ij}a_{jh}a_{hi} + a_{ih}a_{hj}a_{ji}]$$

$$= \frac{1}{2}[A_{(i,:)}AA_{(:,i)} + A_{(i,:)}^T A^T A_{(:,i)}^T] = A_{(i,:)}AA_{(:,i)} = A_{ii}^3 \tag{4.17}$$

where we symbolize with $A_{(:,i)}$, $A_{(i,:)}$ the $i-th$ column and line of A correspondingly. The same notation is employed for A^T. Similarly, for the triangles belonging to the category middleman, we have:

$$t_i^4 = \frac{1}{2}\sum_{j,h}[a_{ij}a_{hj}a_{hi} + a_{ih}a_{jh}a_{ji}]$$

$$= \frac{1}{2}[A_{(i,:)}^T AA_{(:,i)}^T + A_{(i,:)}A^T A_{(:,i)}] = A_{(i,:)}A^T A_{(:,i)} = (AA^T A)_{(ii)} \qquad (4.18)$$

For the categories 2, 3 we compute that:

$$t_i^2 = \frac{1}{2}\sum_{j,h}[a_{ji}a_{jh}a_{hi} + a_{ji}a_{hj}a_{hi}]$$

$$= \frac{1}{2}[A_{(i,:)}^T AA_{(:,i)} + A_{(i,:)}^T AA_{(:,i)}] = (A^T A^2)_{ii} \qquad (4.19)$$

$$t_i^3 = \frac{1}{2}\sum_{j,h}[a_{ij}a_{jh}a_{ih} + a_{ij}a_{hj}a_{ih}]$$

$$= \frac{1}{2}[A_{(i,:)}AA_{(:,i)}^T + A_{(i,:)}AA_{(:,i)}^T] = (A^2 A^T)_{ii} \qquad (4.20)$$

For weighted networks, the adjacency matrix A is replaced by $W^{\frac{1}{3}}$. In the case that we want to compute the participation of triangles belonging to the category/pattern p among all formed triangles, we can divide the actual number of triangles of the category p with the number of all formed directed triangles, i.e., $\frac{t_i^p}{t_i^D}$. Finally, for computing the network-wide clustering coefficient for each one pattern, we average over all the network the corresponding local clustering coefficients of all the nodes, similarly to the case of binary and weighted network graphs.

4.5   Centrality

In various network types, it has been required/desired to characterize the importance of the involved network elements (network nodes), both individually and with respect to the overall network structure. Centrality has been conceived as an evaluation metric for characterizing this aspect of networks. Typically, the focus is on the importance of nodes or connection links, but other features of the graph structure under consideration may also be considered. The "centrality" metric of a node or link is a direct measure of its importance in various networking operations or applications. In addition, a "centrality" metric characterizing the network cumulatively is also conceived as an overall metric characterizing the expected significance of each node in the network, on average. Such metric of importance (i.e., centrality) could be subjective, depending on numerous aspects of a network, such as the structure, the network objectives, network operation, and even other more context-oriented factors characterizing a network. For these reasons, various centrality

Table 4.1: Summary of centrality types in complex and social network analysis.

Type	Metric	Computation
Degree	node degree	distributed
Closeness (proximity)	node pair distance	centralized
Betweenness	shortest path	centralized/
		approximations
Eigenvector	adjacency matrix eigenvalues	distributed

definitions have been established and employed in social and complex communication networks. In the following, we will provide the most important ones and then provide some approaches for effectively computing them. Table 4.1 summarizes the types of centrality measures that will be presented in the sequel of this chapter. Perhaps the most prominent ones are the betweenness-based centrality metric and the eigenvector-based centrality, both of which could be tough to compute in an efficient and distributed manner, but also both bearing the most useful information on node/link importance in networks from various perspectives that will become more evident in the sequel.

For the majority of centrality metrics there are essentially two equivalent values defined, one absolute and one relative, the second defined with respect to the maximum possible achieved value of each centrality metric in each case. In addition, the definition of centrality is inherently focused on individual users. However, the centrality values of each individual node/user can be used in order to obtain a macroscopic definition of graph centrality, as explained previously. In the following, we mainly focus on node centralities and briefly mention how graph centrality may be obtained in each case of a specific centrality metric, when applicable.

4.5.1 Degree Centrality

Motivated by the simple observation that in a star topology the most "central" and thus the most important node for the sustainment of the network is the middle node (Figure 2.4), the simplest and perhaps the most intuitive concept of centrality is that it should be some function of node degree, since in the star topology the distinctive difference of the middle node compared with the rest is the node degree. In this case, degree centrality, namely the centrality measure defined as a function of node degrees and sometimes referred to as point centrality as well, describes the potential of a node to control the information flow in the network through popularity, i.e., a node with high degree is expected to be able to control a greater portion of information flowing in a network, compared to a less popular (lower degree) node.

According to the previous discussion, degree centrality can be a function of the value of node degree for each node. The simplest of these functions is a linear function of the node degree value, and regarding the overall network centrality, a linear function of the average node degree can be defined as the network degree centrality. Assuming that $A = [a_{ij}]$ is the adjacency matrix of a network topology, then the degree centrality and relative degree centrality [38, 69, 71, 103, 119] of a node k can be obtained as:

$$C_D(k) = \sum_{i=1}^{n} a_{ik} \tag{4.21}$$

$$C'_D = \frac{\sum_{i=1}^{n} a_{ik}}{n-1} \tag{4.22}$$

For a vertex k completely isolated from the rest of the node vertices, $C_D(k) = 0$. The magnitude of $C_D(k)$ is partly a function of the actual size of the network for which it is calculated. Depending on the application framework, an absolute value of centrality, such as $C_D(k)$, would work smoothly. However, in some cases a measure that is independent of network size is required. The relative centrality metric C'_D is defined for this purpose.

The degree centrality was the first metric to be used to assess the importance of network nodes. The earliest works on vertex centrality emerged as early as the mid 1950s ([23]), initially for social network analysis. Later the concept of vertex centrality was extended in communications networks as well. Vertex centrality is convenient in the sense that in this way, the degree distribution of the network provides a holistic centrality picture for the corresponding graph.

In general, degree centrality has not been employed extensively, and in fact, it is a measure that has received little attention. The main reason is that it does not capture the dynamics of information flow in a network, and thus the centrality values yielded are not as handy as those yielded by other metrics. For instance, a node might indeed have a high value of node degree; however, for a number of reasons, information might flow through paths that do not include this specific node, e.g., because for security purposes it is safer to go through less popular nodes, etc. Degree centrality appears to be more suitable for applications such as malware propagation, where the degree of a node significantly affects the dynamics of infections and it is the actual factor determining system evolution. In that case, centrality is suitably captured and nodes with high centrality (i.e., degree) should be better protected from targeted attacks. Similarly, in case of random attacks, the network will be resilient due to the restricted node subset each having high node degree, i.e., centrality. Thus, centrality is successfully captured in this application. However, this is not the case with other networking application areas.

In several other applications of interest, such as routing, QoS provision, delay, and throughput management, node importance cannot be quantified solely on the basis of node degree. For instance, traffic bottlenecks are not certain to appear in nodes with high degrees. On the contrary, depending on environmental conditions, the type of application and network structure, bottlenecks could easily form even to the more loosely connected nodes. For these cases, alternative measures of centrality have been developed, in order to capture more accurately the relative importance of a node with respect to all the involved parameters. The next two subsections provide such attempts, despite the fact that that an absolute centrality measure does not seem to have appeared yet.

4.5.2 Closeness (Path) Centrality

A more involved centrality metric, which is also related to the control of communication, but in a different way, is a measure of the relation-importance of a node in the network that identifies the more spatially central nodes. Spatial metrics are inherently dependent on the definition of employed distance metrics. In graph theory the most frequent distance metric is that of hop count. Thus, a centrality metric based on the spatial features of a network topology is one that assigns high centrality to those nodes that are relatively in a central position in the topology with respect to the edges of the network and the employed distance metrics. By network edges here we mean the more distant nodes in the spatial network distribution as shown in Figure 4.4, where all edge nodes have been denoted by an "E" symbol.

The definition of closeness centrality (alternatively referred to as proximity or path based centrality) is based on the distances of each node from all the

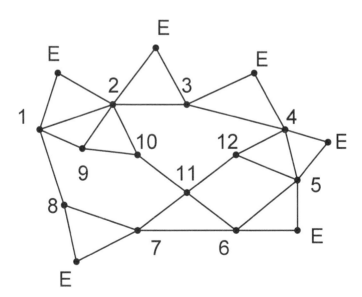

Figure 4.4: Network edge nodes. All nodes denoted by "E" are the edge nodes of the network.

rest. As explained before, the distance in a network graph is meant in the sense of the length of the shortest path from a node to another node given the topology of the network. Denoting by $d(i,k)$ the shortest path distance between nodes i and k, the proximity and relative proximity centralities are [24, 71, 103, 119]:

$$C_P(k) = \left(\sum_{i=1}^{n} d(i,k) \right)^{-1} \tag{4.23}$$

$$C'_P = \left(\frac{\sum_{i=1}^{n} d(i,k)}{n-1} \right)^{-1} = \frac{n-1}{\sum_{i=1}^{n} d(i,k)} \tag{4.24}$$

The closeness centrality metric depends on the closeness (proximity) of a node to the rest of the network nodes. Proximity describes quantitatively the notion of how close to each other nodes are, and the corresponding centrality metric essentially describes how close a node is to the rest of the network nodes. Given the latter, a node that is relatively "close" to most of the rest of the network nodes, for which its proximity based centrality will be high, will be more "central" than a node that is relatively "far" from other nodes in the network, and thus, its expected proximity centrality value will be low. In that sense, a message originating in the most central node in the network would spread throughout the entire network (through a flooding scheme) in minimum time, since the message will follow exactly all the shortest paths from that central node to all the rest and the total length sum of these paths will be the lowest possible. Thus, the most central vertex in the network can be defined as that with the minimum cumulative cost for reaching all the rest. A simple way to quantify this is the one given above in Eq. 4.24, where the centrality of a vertex (absolute measure) is measured by summing the geodesic distances from the specific vertex to all other vertices in the graph and then inverting this value, so that highest proximity centrality corresponds to lowest communication cost (expressed as distance from the rest of the nodes). The calculation of $C_P(k)$ is simple and straightforward, and several matrix methods used in other centrality measure computations that will be described in the sequel can be employed.

The above definition of the absolute value of $C_P(k)$ reveals that its value depends on the order of the network, i.e., the number of network nodes. For this reason, a network size independent metric, namely relative proximity centrality, has also been provided, which averages the value of proximity centrality over the nodes of the network. The relative measure of closeness centrality C'_P takes a unit value when a vertex is maximally close to all other vertices and decreases as the average distance of a specific vertex to all other vertices increases.

Both the degree and proximity based measures of centrality suffer from the fundamental problem that the underlying network has to be fully connected. By definition, in disconnected networks, node degrees could have zero values and shortest paths could have infinite values. Thus, in addition to the fact that these two measures do not inherently take into account information control dynamics, even though proximity centrality does address an aspect of the information flow process—namely information dissemination, they also exhibit quantitative problems, when networks have isolated nodes (i.e., disconnected networks) or network connectivity varies dynamically (intermittently connected network types, such as delay-tolerant networks in communications networks).

In Subsection 4.5.3, the described betweenness-based centrality definition does not suffer from the problems faced by degree and proximity centrality, and could be equally used both for connected and intermittently connected networks.

Approximation of Closeness Centrality

As has become evident when computing vertex proximity centrality measures, it is required to solve the all-pairs shortest-path problem (which could be tackled in $O(nm + n^2 \log n)$ or $O(n^3)$ time, for a graph with m edges and n vertices) and then distribute the information back to the nodes [11]. Of course, each time a topology changes, the whole process needs to be repeated. Given also the fact that modern social and communications networks scale in the order of millions of actors/users, more efficient computation methods are required.

In the following we describe a random sampling technique to approximate closeness based centrality of vertices in a weighted graph, within an additive error of $\epsilon\Delta$ with high probability of time complexity $O(\frac{\log n}{\epsilon^2}(n \log n + m))$, where $\epsilon > 0$ and Δ is the diameter of the graph. Especially for small-world type networks where the diameter scales as $O(\log n)$ rather than as $O(n)$ (see Section 5.2), the following approach provides a near-linear time $(1 + \epsilon)$-approximation for the proximity centrality of all vertices [60].

The approximation algorithm, denoted by RAND, chooses k sample vertices and computes single source shortest paths (SSSP) from each sample vertex to all other vertices in the graph. The whole process takes place in k iterations, in each of which, e.g., iteration i, a vertex v_i is uniformly picked at random and the SSSP problem is solved with v_i as the source. Eventually, the centrality measure of vertex u is given by $\hat{c}_u = \frac{1}{\sum_{i=1}^{k} \frac{n d(v_i, u)}{k(n-1)}}$.

It has been shown in [60] that the expected value of $1/\hat{c}_u$ is equal to $1/c_u$, i.e., $E\left[\frac{1}{\hat{c}_u}\right] = \frac{1}{c_u}$. It is also shown that for a connected graph G with n vertices and diameter Δ, with high probability, algorithm RAND described above computes the closeness centrality estimator $\frac{1}{\hat{c}_u}$ to within $\epsilon\Delta$ of the inverse centrality $\frac{1}{c_u}$ for all vertices u of G, using $\Theta\left(\frac{\log n}{\epsilon^2}\right)$ samples for $\epsilon > 0$.

Taking into account an efficient algorithm for solving the SSSP problem in $O(n \log n + m)$ by Fredman and Tarjan [67], the total running time of RAND is $O(km)$ for unweighted graphs and $O(k(n \log n + m))$ for weighted graphs. Thus, for $k = \Theta \left(\frac{\log n}{\epsilon^2} \right)$ samples, RAND is an $O(\frac{\log n}{\epsilon^2}(n \log n + m))$ approximation algorithm for proximity centrality within an additive error of $\epsilon \Delta$ with high probability.

Other approximation methods have been presented in [166], [73], and [11] in various capacities and covering different objectives. Compared to [60], which focuses on closeness centrality, the other three focus on betweenness-based centrality approximation, which will be extensively presented in the proceeding subsections.

4.5.3 Betweenness Centrality

As already mentioned, typical centrality definitions are not appropriate for use in intermittently connected networks, where at several instances various nodes become disconnected permanently or temporarily (e.g., ad hoc, DTNs, mesh and sensor in communications). In fact, for some of the earliest employed definitions of centrality, where the centrality of a vertex is some function of the sum of all minimum distances between the specific vertex and all other vertices, computing centrality in intermittently connected nodes would not be numerically possible. Thus, alternative centrality definitions are required in order to ensure transparent and meaningful representation of the concept. Such definitions should be able to handle such degenerate cases (disconnected nodes) and yield the same results as the previous ones (at least relatively, in the sense that nodes of higher centrality, according to one definition, retain high centrality according to another definition). The appropriate definition should also be able to segregate better between different classes-groups of nodes with respect to their centrality metric.

Such a centrality metric could again be based on the notion of shortest paths and more specifically on the frequency with which a node falls on the shortest paths connecting other pairs of vertices of the underlying network graph. This definition illustrates the idea that a node lying with high frequency on emerging shortest paths has more potential to control the actual information flow through the network, and thus should be considered more central than others. Intuitively, in a communication network where routing of information usually takes place along the shortest paths (in any manner they are determined), a node with high frequency of falling on such emerging shortest paths has more potential to withhold/distort/impact the transmitted information and thus it has a more central role in the process of information communication. Thus, it is the potential for control that determines the centrality of vertices in this case. The term betweenness has been employed to quantify the aforementioned frequency of a node to fall on the shortest paths of other nodes [68, 70, 103, 119].

The definition of betweenness centrality is straightforward when only one geodesic (shortest path) connects each pair of nodes. However, when there are several geodesics connecting a pair of points (alternative paths, all of the same 'shortest' length), the situation becomes more complicated. Then, a vertex that falls on some but not all of the geodesics connecting a pair of others has a more limited potential for control. Such partial potential of nodes to control information flow is reflected through the notion of partial vertex betweenness. Given a vertex k and an unordered pair of vertices $\{i, j\}$, where $i \neq j \neq k$, the partial betweenness $b_{ij}(k)$ of node k with respect to the pair (i, j) can be defined in the following way. If i and j are not reachable from each other, namely there exists no path connecting them at all, then $b_{ij}(k) = 0$. If nodes i and j are reachable from each other, i.e., there exists one or more geodesic paths connecting them, then the probability of using any one of the alternative geodesics between i and j is $\frac{1}{g_{ij}}$, where g_{ij} is the number of geodesics connecting i and j. This is also the probability that a message passes along any particular geodesic among alternatives between i and j. The potential for k to control information flowing between i and j can be defined as the probability that k falls on a randomly selected geodesic connecting i and j. If $g_{ij}(k)$ is the number of geodesics connecting i and j also containing k, then the aforementioned probability is:

$$b_{ij}(k) = \frac{g_{ij}}{g_{ij}(k)} \tag{4.25}$$

In the event that k is on the only geodesic between i and j, or on all available geodesics between i and j, then $b_{ij}(k) = 1$.

To determine the overall betweenness centrality of vertex k, we sum the partial betweenness of k for all unordered pairs of vertices $i \neq j \neq k$:

$$C_B(k) = \sum_{i \neq j \neq k, i < j}^{n} \sum^{n} b_{ij}(k) \tag{4.26}$$

where n is the number of vertices in the graph. Whenever there are alternative geodesics between node pairs, the betweenness centrality $C_B(k)$ of k is increased in proportion to the frequency of occurrence of k among those alternatives, which means that multiple available geodesics are indeed taken into account. This allows computing a centrality value even for a disconnected node, i.e., zero centrality, indicating that such a node correctly bears the lowest possible centrality value.

Freeman (1977) [68] proved that the maximum betweenness centrality of any vertex in a graph is achieved only by a central vertex in a star graph (or of a wheel graph) and equals

$$C_B^{\max}(k) = \frac{n(n-1)}{2} - (n-1) = \frac{n^2 - 3n + 2}{2}. \tag{4.27}$$

A constructive proof for the star graph having the node with maximum centrality can be found in [68]. Therefore a relative centrality measure for a vertex, which will be independent of the size of the network over which it is calculated, is:

$$C'_B(k) = \frac{2C_B(k)}{n^2 - 3n + 2} \qquad (4.28)$$

The above definition indicates the major drawback of the betweeness-based centrality. It requires the computation of the total number of shortest paths in a network by each node, thus making the computation of $C_B(k)$ a centralized process. This could be very demanding, especially for large scale networks with thousands of nodes (sometimes millions as in online social networks). Matrix methods have been developed and more discussion on such algebraic approaches will be provided in the subsection 4.5.5.

However, the inherent remaining difficulties and centralized processing even of algebraic approaches have stirred significant efforts for distributed computations of centrality, most of which have yielded several acceptable approximation mechanisms. This is exactly the topic of the following subsection and some popular approximation efforts will be described in the sequel.

4.5.4   Betweenness Centrality Approximation Methods

The previous subsections defined betweenness centrality and explained its importance in social networks analysis. However, as the scale of the networks analyzed increases rapidly and in most cases reaching the orders of millions of nodes and even more edges, e.g., online social networks, mobile cellular networks, global market players, etc., exact computation of betweenness centrality becomes prohibitive, since all shortest paths in the network and all shortest paths passing through each node need to be computed. Thus, a strong motivation exists to devise methods for computing betweenness centrality values in networks of millions of nodes within a few minutes. Within this context, reasonable approximation methods are required that are capable of approximating within a reasonable error constraint the betweenness centrality value of all network nodes, and consequently enabling segregation of network nodes according to their centrality importance.

Brande's Computation Approaches

Brandes [40] presented an exact algorithm for computing betweenness centrality of all nodes in a distributed fashion, which was based on solving a Single Source Shortest Path (SSSP) problem from each node. An SSSP procedure yields a Directed Acyclic Graph (DAG) with all shortest paths emanating from each node. By backward aggregation of properly defined counter values across these paths, the contributions of the paths on betweenness counters can be computed in linear time. Depending on each topology the exact distributed algorithm takes time $\Theta(nm)$ over unweighted graphs to $\Theta(nm + n^2 \log n)$ for arbitrary weighted graphs, where n is the number of nodes and m the number of edges. However, even though this is a polynomial time approach, it remains prohibitive for very large networks in the order of millions. A massively parallel implementation of the exact Brandes' approach by Bader and Madduri [11] can handle a few million nodes. Nevertheless, the need for approximation approaches remains.

The exact centrality computation approach is based on the notion of dependency of a source vertex $s \in V$ on a vertex $v \in V$, defined as:

$$\delta_{s*}(v) = \sum_{t \neq s \neq v \in V} \delta_{st}(v) \tag{4.29}$$

where $\delta_{st}(V)$ is the fraction of SPs between s and t that pass through vertex v.

Using this quantity, the betweenness score of node v can be expressed as:

$$BC(v) = \sum_{s \neq v \in V} \delta_{s*}(v) \tag{4.30}$$

Let $P_s(v)$ denote the set of predecessors of a vertex v on shortest paths from s, i.e., $P_s(v) = \{u \in V : (u, v) \in E, d(s, v) = d(s, u) + w_{uv}\}$, where w_{uv} is the weight of edge (u, v).

Brandes showed that the dependencies satisfy the following recursive relation:

$$\delta_{s*}(v) = \sum_{w : v \in P_s(w)} \frac{g_{sv}}{g_{sw}} (1 + \delta_{s*}(w)) \tag{4.31}$$

First, n SSSP computations take place, one for each $s \in V$. The predecessor set $P_s(v)$ are updated in the above computations. Then for every $s \in V$, the dependencies $\delta_{s*}(v)$ for all other $v \in V$ is computed by utilizing information from the shortest path trees and predecessor sets. The sum of all dependency values eventually yields the centrality value of a vertex v. The $O(n^2)$ space requirements can eventually be reduced to $O(n + m)$.

Extending the exact approach, Brande and Pich [41] turn it into an approximation algorithm based on sampling. Starting with only a subset of k starting nodes, denoted by *pivots*, they use the same backwards aggregation strategy as in the exact approach. A random sample of k starting nodes (pivots) turns out to work well, yielding an unbiased estimator of betweenness, namely that the expectation of the estimated betweenness is the actual betweenness. However, it produces large overestimates for unimportant nodes lying topologically near a pivot.

Bisection based Approximation Approach

In this approach, the main concept is to reduce the contributions of nodes close to the pivot nodes, which as explained above essentially cause overestimating centrality values for these nodes. A general framework for betweenness approximation taking advantage of Brande's random sampling approach enables two implementations, one where the contribution of a sample depends linearly on the distance of the sample (termed linear scaling sampling approach) and a second one where the sample contributes on the second half of a path (referred to as bisection scaling approach).

We define a length function $\ell : E \to \mathbb{R}$ on the edges and a scaling function $f : [0,1] \to [0,1]$. For a path $P = \langle e_1, ..., e_k \rangle$ the length of the path is given by $\ell(P) := \sum_{1 \leq i \leq k} \ell(e_i)$. In this approach, in each iteration, one of $2n$ possible path searches is performed with uniform probability $1/2n$ (forward or backward search from a pivot). The scaled contribution of a path is defined as

$$\delta_P(v) := \begin{cases} \frac{f(\ell(Q)\ell(P))}{g_{st}}, & \text{for a forward search} \\ \frac{1-f(\ell(Q)\ell(P))}{g_{st}}, & \text{for a backward search} \end{cases}$$

Based on that, node v obtains a contribution of

$$\delta(v) := \delta_s(v) := \sum_{t \in v} \sum \{\delta_P(v) : P \in SP_{st}(v)\} \tag{4.32}$$

for a forward search and

$$\delta(v) := \delta_s(v) := \sum_{s \in v} \sum \{\delta_P(v) : P \in SP_{st}(v)\} \tag{4.33}$$

for a backward search.

It has been shown in [73] that $X := 2n\delta(v)$ is an unbiased estimator for the centrality of node v, namely that $E[x] = C_B(v)$. In order to obtain an unbiased estimator of the betweenness of v, it suffices to average k independent values, i.e., $\frac{X_1 + ... + X_k}{k}$.

For a constant function $f(x) = 1/2$ the Brandes' algorithm is obtained, while for linear scaling $f(x) = x$ and $f(x) = \begin{cases} 0, & \text{for } x \in [0, 1/2) \\ 1, & \text{for } x \in [1/2, 1] \end{cases}$ for bisection scaling. Bisection scaling seems to be more successful in reducing the contributions for nodes close to pivots; however, linear scaling is easier to implement.

In order to compute g_{st} on the fly while traversing a shortest path, a $g_{ss} = 1$ and for $s \neq t$, $g_{st} = \sum_{v \in pred(t)} g_{sv}$, where $pred(t)$ is a set containing the immediate predecessors of t in the shortest path DAG. In a subsequent aggregation phase, the nodes are processed in reverse topological order, namely by non-increasing distance from s. Thus,

$$\delta_s(v) = \sum_{w \in succ(v)} \frac{g_{sv}}{g_{sw}} (1 + \delta_s(w)) \tag{4.34}$$

where $succ(v)$ denotes the immediate successors of v in the shortest path DAG.

Linear scaling can be implemented by using the original edge weights as the length function ℓ, and with only minor deviations from Brandes' aglorithm:

$$\delta_s(v) = \sum_{w \in succ(v)} \frac{\mu(s, v)}{\mu(s, w)} \frac{g_{sv}}{g_{sw}} (1 + \delta_s(w)) \tag{4.35}$$

where $\mu(s,w)$ is the shortest path distance from s to w.

In the bisection scaling implementation, unit distances are employed for the length function ℓ. Aggregation follows a depth-first traversal of the shortest-path tree, which allows storing the path from s to the explored node in a simple array. The difference compared to Brandes' algorithm is that when a node v at depth d is explored, the current value $\delta_s(v')$ of node v' on position $\max\{0, \lfloor d/2 \rfloor - 1\}$ is decremented, which has the effect of dropping the contribution of v from the aggregation where it is prescribed by the scaling function f.

In a variant called *bisection sampling* a parent pointer is randomly sampled for each node t in the shortest path DAG. In this case, if a parent p of w is selected with probability $\frac{g_s p}{g_s w}$, an unbiased estimator is obtained. In addition, without disturbing the unbiased estimator, more information can be extracted out of the shortest path DAG by performing several sampling steps for the same DAG.

Sampled bisection sampling yields good approximation of betweenness for less important nodes as well that already have a small number of pivots.

The variations of Brande's algorithm need slightly more time than Brande's algorithm for evaluating the shortest path DAG, while the same approximation quality is obtained and works well even for huge networks. However, for some directed networks, none of the approximation algorithms give very convincing results, since a good approximation requires time almost as much as an exact calculation.

Adaptive Sampling Betweenness Vertex Centrality Approximation

One of the problems identified in the above approach for approximating betweenness centrality is that the approximation values computed are dependent on the vertices from which the shortest path computations are initiated, i.e., the pivot, sometimes referred to as pivot nodes. It has been observed that a random selection of source vertices to start the computations is superior to deterministic strategies.

Most of the approximation methods follow the approach of computing concurrently the centrality scores of all vertices in a topology, which essentially mandates solving at least as many SSSP problems as the vertices in the graph. However, the complete centrality scores of a topology are not always required. In many applications and cases, it is only required to know the centrality score of a single vertex, and at some times only the relative centrality value of the vertex compared to the centrality scores of the rest of the network vertices, i.e., how the centrality value of a node compares in an ordering of all nodes' centrality values. This should take place in faster time than the approximation of centrality of all nodes and should also require less resources (produce less overhead as well).

In the sequel, an adaptive sampling based algorithm is presented, which addresses the above considerations for approximating the centrality of selected network vertices and it is based on the adaptive sampling technique.

The adaptive sampling centrality approximation of a given vertex estimates the centrality value by sampling a subset of vertices and performing SSSP computations from these vertices. The number of samples required varies with the information obtained from each sample, hence the term *adaptive*.

The first step deals with the degenerate case, where the neighbors of the given vertex v induce a clique. In this case, all these neighbors will communicate through their direct links and thus vertex v will not lie on any shortest path. Thus, the betweenness centrality value of such nodes v will be zero.

The adaptive sampling algorithm proceeds as follows. It repeatedly samples a vertex $v_i \in V$ and computes the SSSP from v_i using either Breadth-first-search or Dijkstra's approach for computing shortest paths. For each SSSP computation it maintains a running sum S of the dependency scores $\delta_{v_i^*}(v)$. The sampling continues until S is greater than cn for some constant $c \geq 2$. Assuming the total number of samples required was k, the estimated betweenness centrality score of node v is given by $C_B(v) = \frac{nS}{k}$. The adaptiveness of the algorithm lies in the fact that the number of samples required varies for different topologies and types of networks. Feedback is acquired by the running accumulated value of the dependency sum, based on which it is determined whether the samples taken are sufficient or not.

In [11] a lower bound on the expected number of samples required before stopping is provided. It is shown that for $0 < \epsilon < 0.5$, with probability greater than $1 - 2\epsilon$, the above algorithm estimates the centrality of a given vertex within a factor of $1/\epsilon$. The results have been shown to hold for both high centrality nodes and the ones with lower values of centrality, even nodes with very low ones (the latter shown through simulations in [11]). More specifically, the authors have shown through experiments on real-world graph traces that the error for random graphs underlying a network is about 5% and roughly about 10% for the rest of the considered network types.

Constrained Centrality Approximation

In some applications, it is required to find the node with the maximum centrality value (or a relative ordering of node centrality values in order to find the top centrality ranking nodes) given specific constraints. Such an example is the gateway design problem in [166]. This is a dynamic mesh network construction application, where given a connected backbone network $G(V, E)$ and an initiator node $i \in V$, it is required to find distributively the node v with maximum centrality $C(v)$ such that the least transmission overhead is imposed (total number of messages exchanged) at the minimum possible time (time elapsed).

FACE algorithm [166] addresses the above problem by decomposing it in two sub-problems, namely centrality computation and extrema finding. It is focused on the computation of closeness centrality rather than betweenness centrality; however, we present it here since the method utilizes the random sampling SSSP approach presented above for approximating betwenness centrality. For the first, a spanning tree from each node is built, and all the distances from each root node to all the other nodes along the spanning trees are collected by each root (similarly to and exploiting a proactive link state routing protocol). The spanning tree of the initiator node i is denoted as *primary spanning tree* (PST) and it is used mainly for finding the extrema and reporting it to the iniator, while the rest are denoted by secondary spanning trees (SST) and are mainly used for centrality measuring.

A flooding based spanning tree construction approach is employed, where a root sends messages to all its neighbors and all the neighbors propagate them to their neighbors in turn, while keeping track through a specific field in each message. This approach adopts the random sampling centrality approximation technique and further utilizes the observation that if a spanning tree is constructed in a network, its leaves are highly likely to reside on the perimeter of the actual underlying topology. Thus, such close-to-the-border leaves are good pivot candidates, since centrality measurement is largely related to the distances of nodes to the border of the topology. FACE uses exactly the root and leaves of a spanning tree as the sample points (pivots) for centrality approximation. The approach essentially computes the relevant centrality values of the network nodes.

The maximum centrality node is found on the basis of the PST after the centrality values have been approximated. Starting at the PST leaves, each reports to its parent its centrality value. A non-leaf node, upon receipt of the centrality value of its children, compares its own centrality value with the maximum reported value from its children and sends the largest to its parent. Consequently, the initiator can determine which node has the maximum centrality.

4.5.5 Eigenvector Centrality

The eigenvector centrality [35, 36, 38, 71, 103, 129] is defined as the principal eigenvector of the adjacency matrix A (Section 2.1.8), representing the network graph under consideration. The principal eigenvector of a matrix corresponds to its largest eigenvalue. The difference between eigenvector centrality and the already presented centrality measures lies in the fact that eigenvector centrality takes into consideration the importance or centrality of the direct neighbors of a node. This means that a node has a high eigenvector score if it is adjacent to nodes having themselves high eigenvector centralities. This is clear from the mathematical formula providing the eigenvector scores (which coincides with the eigenvectors' computation).

If λ is the maximum eigenvalue of A and v the corresponding eigenvector (column vector), then

$$\lambda v = Av \Rightarrow$$
$$v = \frac{1}{\lambda} Av \tag{4.36}$$

If the network has N nodes then A is a $N \times N$ matrix and $v = [v_1 \ v_2 \ v_3 ... v_N]^T$. Then

$$v_i = \frac{1}{\lambda} \sum_{j=1}^{N} a_{ij} v_j, \ \forall \ i \tag{4.37}$$

We also note that if A is a connected graph then it is irreducible and according to the Perron–Frobenius Theorem (Section 2.1.8), its largest eigenvalue is always positive. In this case also, each $v_i > 0$, $\forall \ i$. Similar definition of the eigenvector centrality can be given for weighted network graphs where A is replaced by the weight matrix W. It is important to note that Google's algorithm for ranking Web pages, i.e., PageRank, is based on a technique similar to the one for the computation of the eigenvector centrality [121, 129].

4.5.6   Example of Centralities' Computation

In this subsection, we provide a handy example depicting the computation of centralities of the network shown in Figure 4.5.

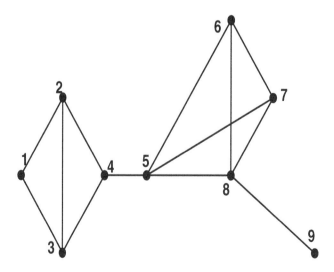

Figure 4.5: Network topology used for the computation of node centralities.

Table 4.2: Ranking of node importance, i.e., centrality (and corresponding centrality value) with respect to each one centrality metric presented in this chapter.

Degree Centrality		Closeness (Path) Centrality		Betweenness Centrality		Eigenvector Centrality	
Node IDs	C_D	Node IDs	C_P	Node IDs	C_B	Node IDs	v
8, 5	4	5	0.0769	5	0.7778	5	0.498
2, 3, 4, 6, 7	3	4	0.0714	4	0.6944	8	0.472
1	2	8	0.0625	8	0.3611	6, 7	0.438
9	1	7, 6	0.0588	2	0.1667	4	0.0.254
		2, 3	0.0556	1, 3, 6, 7, 9	0	2, 3	0.159
		9	0.0435			9	0.147
		1	0.0417			1	0.099

Table 4.2 presents the ranking of nodes in decreasing order and their corresponding centrality values for each one centrality metric, i.e., Degree Centrality (C_D), Closeness (Path) Centrality (C_P), Betweenness Centrality (C_B), Eigenvector Centrality (v). We observe that the ranking differs with respect to the centrality measure used. However, all the centrality measures agree that the node with ID 5 is the most important. Degree centrality, as expected, is shown to rank in higher levels nodes with more direct neighbors. Closeness centrality, due to its mathematical form, assigns higher values to nodes 5 and 4, which are located at the center of the topology and therefore in close hop distance from all others. Betweenness centrality considers as important only nodes participating in the shortest paths between other node pairs, i.e., nodes with IDs 5, 4, 8, 2, assigning 0 centrality to nodes at the edge of the topology. Finally, the Eigenvector centrality takes into account the centralities of the neighbors of a node in the computation of its centrality value, and therefore, a node with important neighbors is itself important (has high centrality).

4.6   Prestige

From the previous presentation of centrality, it has probably become apparent that in the consideration of centrality, no directional properties of the developing relations have been taken into account. Namely, the underlying topologies were considered undirected, and the centrality was mainly determined on the basis of geodesic distance (betweenness, closeness) or popularity (degree). The interest was in the existence of a relation (connectivity and minimum distance) signifying access or control to resources and information flow, rather than on the direction in which this access or control takes place. However, in many technological and social applications, the direction in which the above relations take place is important and needs to be quantified. Prestige is a measure coined to express exactly this requirement. Thus, prestige may be considered as a centrality measure for directed networks. The difference is that prestige measures, compared to centrality measures, express not only importance, but also a non-reciprocated direction of increasing importance. If this direction can be reciprocated, centrality measures suffice. Such an example is persons who are elected to board bodies and thus obtain prestige in their social communities. The selection cannot be reciprocated (since it is determined by legal vote) and thus a directed network is required to denote such relations. The prestige is more appropriate for quantifying importance in this case. Consequently, prestige is a more refined concept, signifying in addition the direction to which a tie increases importance.

Based on the above intuitive definition, prestige should increase when a node becomes the object of more relations (in-coming edges), but not necessarily when the node itself initiates the ties (out-going edges). A directed network may be used to determine relations received (in-edges) and initiated (out-edges). Prestige measures focus on relations received (both direct and indirect). In that sense, centrality measures can also be used for relations initiated (both direct and indirect), which signify the degree of control that these nodes have on the flow of information and access to resources of the network under analysis. Degree and closeness (proximity) centrality are easy to apply to directional relations while betweenness is not because of its reliance on non-directed paths.

In the following, we present some of the prestige measures that express the above concepts and note that research on prestige is relatively new compared to the rest of complex/social network analysis metrics, which leaves a significant margin for new research, as will also be explained in the last chapter of the book.

4.6.1 Degree Prestige

The simplest measure of prestige is perhaps, as in the centrality case, the degree of a node, and more specifically the in-degree of a node. Since it quantifies the incoming edges of a node, which in general represent "received" behavior by the specific node from its peers in the network, it is a direct measure of the "positive influence" received by the node, which in many situations increases the importance, i.e., prestige, of the specific node.

4.6.2 Influence Domain

Several efforts have been made to extend prestige to indirect relations, contrary to the direct ones described by the above definition of degree prestige. A first extension is to employ the *influence domain* of a vertex t, which is defined as the number of nodes that have a directed 1-hop or multi-hop path towards the sink vertex t. All these nodes essentially represent actors that could potentially influence directly or indirectly the sink node t. The proportion of vertices that belong in the influence domain is computed by dividing the number of vertices in the influence domain by $n-1$ (number of all other vertices). This measure of prestige is meaningful only if the network is not strongly connected, namely if it is weakly connected. For strongly connected networks, closeness (proximity) centrality is a more suitable metric to use.

As a measure of prestige, the influence domain of a vertex does not distinguish between direct and indirect choices, which is not always completely desirable. Usually, direct relations are considered more prestigious than indirect ones. A relation contributes less to prestige if it is mediated by a longer chain of intermediaries. To overcome such drawbacks and allow direct choices to contribute more to the computation of prestige than indirect choices, one can weight each choice by its path distance to the selected vertex. A higher distance will yield a lower contribution to the prestige of another vertex. A simple computation of the mean distance of a vertex from vertices in its influence domain will suffice in this case. However, in order to obtain a more accurate estimation of prestige the average distances must be combined with the size of influence domains.

4.6.3   Proximity Prestige

The actor and group-level prestige indices on proximity or graph distances to each actor can be useful. Actors are judged to be prestigious based on how close the other actors in the set of actors are to them. However, one should also consider the prestige of actors that are proximate to the actor in question. If many prestigious actors choose an actor, then that should be given more weight than if many non-prestigious actors choose an actor. This example motivates another prestige measure called proximity prestige. The latter would constitute a measure of how close other actors are to a given actor.

Proximity prestige is another measure of prestige based on the closeness notion of centrality, extended to directed networks. The proximity prestige of a selected vertex is computed by dividing the influence domain of this vertex (expressed as a ratio) by the average distance from all vertices in the influence domain. A larger influence domain and a smaller distance yield a higher proximity prestige score. The maximum proximity prestige score is achieved if a vertex is directly linked (selected) by all other vertices, i.e., when all other vertices create a directed connection to the specific vertex achieving the maximum proximity prestige score. In that case, the proportion of vertices in the influence domain is 1 and the mean distance from these vertices is 1 as well, yielding a proximity prestige of 1. Vertices without influence domain receive minimum proximity prestige, which is equal to zero.

A direct outcome of the above definition of proximity prestige is that if a network is strongly connected, then the proximity prestige equals the input closeness centrality.

Another two measures of prestige may be considered two distinct types of vertices: hubs and sinks (sometimes referred to as authorities). The latter describes a node where other nodes point at it, while the first describes nodes that point to some other nodes. A good hub is a node pointing to many good sinks, while it is a good sink if it points to many good hubs. Hubs and sinks are characterized based on weight values computed for each node, which in turn was computed by solving the eigenvector problem AA^T for hubs and $A^T A$ for sinks, where A is the adjacency matrix of a network.

As expected from the above definitions and analyses, most prestige measures are very similar in the computed score values, except for betwenneess centrality, which measures prestige in a quite different way [170].

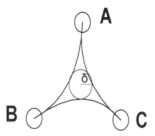

Figure 4.6: δ-thin triangle.

4.7   Curvature

In this section, we are going to study a geometric property of complex networks denoted as negative curvature or hyperbolicity. This intrinsic geometrical feature exists in many complex networks, and it is interconnected with their topological characteristics and affects their performance, as will be more clearly explained shortly. In the following, we will explain and discuss the definition of the network curvature and we will provide brief examples concerning the computation of curvature in some complex networks of interest.

To begin with, let us suppose that we have a geodesic metric space, where d_{ij} is the length of the geodesic between the nodes i, j of the graph G.

Definition 16 *(Gromov's definition of δ-hyperbolicity [114])*
For three nodes, i, j, k, the geodesics d_{ij}, d_{jk}, d_{ki} are constructed. Let the set of nodes on each corresponding geodesic be symbolized as (i, j), (j, k), (k, i). Then we consider a fourth node m and denote the shortest distance between m and all the nodes (i, j) as d_{ij}^m. Similarly, for d_{jk}^m, d_{ki}^m. It is defined as the distance $D_{ijk}^m = \max\{d_{ij}^m,\ d_{jk}^m,\ d_{ki}^m\}$. Then if $\max_{i,j,k} \min_m D_{ijk}^m = \delta$ is finite, then the graph is said to have negative or hyperbolic curvature.

In other words, Definition 16 states that there is a (minimal) value $\delta \leq 0$, such that for any three nodes of the graph connected to each other by geodesics, each geodesic is within the δ-neighborhood of the union of the other two (Figure 4.6).

As can be observed from Definition 16, curvature is a global network feature, contrary to the aforementioned local network features such as degree distribution, strength, local clustering coefficient, and locally computed centralities. Therefore, it can be related to the global performance of the network. To verify this, through numerical evaluation, it can be observed that the load C at the center of a hyperbolic graph scales with the number of nodes N in the graph as $C \sim N^2$ [114], i.e., faster than in a Euclidean Network, where $C \sim N^{1.5}$. It can be stated that the core congestion in networks with negative (or hyperbolic) curvature is worse than that of Euclidean networks, i.e., in hyperbolic networks all the traffic of the network passes through a small core of the network graph. It is important to note that the nodes inside the core are not necessarily high-degree nodes.

At this point, we will refer to two important complex networks and characterize their curvature through numerical studies. In [114], the parameter δ (Definition 16) is numerically computed and plotted for the Watts and Strogatz Model [162] and the Barabási–Albert Model [6], both of which will be presented in detail in the next chapter (Chapter 5). It is shown that δ is finite for the Barabási–Albert Model, which gives rise to scale-free network graphs, but it is infinite for the Watts and Strogatz Model, which produces small-world graphs. Therefore, we can state that the scale-free networks emerge with negative curvature while the small-world networks are flat. The negative curvature of scale-free networks is also shown analytically in [125]. In [125], a two-way procedure is described showing the negative curvature of scale-free graphs. In the first direction, it is proved that a hyperbolic random network is scale-free, by examining its distribution, while in the second direction a scale-free network is proved to have negative curvature.

Since it is complex to compute the network curvature through Definition 16, which however holds for all network graphs, we provide another definition for the special case of planar graphs (Chapter 2). For planar graphs, the curvature can be computed in node-level, using only the local structure of the network. We remind the reader that k_i is the degree of node i. Also, F_{ij}, $1 \leq j \leq k_i$ are the faces incident to i and $d(F_{ij})$ the number of sides of each face. $|V|$, $|E|$, F are the number of vertices, edges, and faces of the graph, correspondingly.

Definition 17 *(Higuchi's curvature for planar graphs) For each vertex i of a planar graph, we define its curvature as $K(i) = 2\pi \left\{ 1 - \frac{k_i}{2} + \sum_{j=1}^{k_i} \frac{1}{d(F_{ij})} \right\}$.*

Theorem 57 *(Higuchi) Let G be a planar graph, if $K(i) < 0$ for every vertex $i \in G$, then G is hyperbolic.*

If $K(i) < 0$ (negative curvature or hyperbolic), we say that i has an angle excess and if $K(i) > 0$, we say that i has an angle defect. The sum $\sum_i K(i)$ is the total curvature of the planar graph. By the Gauss–Bonnet theorem [45, 92], the graph curvature is equal to $K(G) = \sum_{i \in G} K(i) = 2\pi\chi(G)$, where $\chi(G)$ is the Euler characteristic of the graph. For planar graphs, $\chi(G)$ is defined as $\chi(G) = |V| - |E| + F$. Definition 17 is extended for non-planar graphs after their embedding to a suitable torus surface, which however is not easily computable and is out of the scope of the book. In the following, we show an example of curvature computation for a planar graph. For the planar network graph example in Figure 4.7, by using Definition 17, we compute the following, considering that the exterior surface does not count as a face (or else it is a face with ∞ sides).

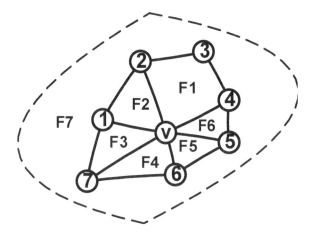

Figure 4.7: Paradigm of curvature computation of a planar graph.

$$K(v) = 2\pi(1 - \frac{6}{2} + 5\frac{1}{3} + \frac{1}{4}) = -\frac{\pi}{6} \qquad (4.38)$$

$$K(3) = 2\pi(1 - \frac{2}{2} + \frac{1}{\infty} + \frac{1}{4}) = \frac{\pi}{2} \qquad (4.39)$$

$$K(2) = K(4) = 2\pi(1 - \frac{3}{2} + \frac{1}{3} + \frac{1}{\infty} + \frac{1}{4}) = \frac{\pi}{6} \qquad (4.40)$$

$$K(1) = K(7) = K(6) = K(5) = 2\pi(1 - \frac{3}{2} + 2\frac{1}{3} + \frac{1}{\infty}) = \frac{\pi}{3} \qquad (4.41)$$

$$K(G) = -\frac{\pi}{6} + \frac{\pi}{2} + 2 * \frac{\pi}{6} + 4 * \frac{\pi}{3} = 2\pi \qquad (4.42)$$

Furthermore, following the Gauss–Bonnet theorem, $\chi(G) = |V| - |E| + F = 8 - 13 + 6 = 1$, leading to $K(G) = 2\pi\chi(G) = 2\pi$.

4.8   Metrics at a Glance

The following Table 4.3, provides a quick overview of the presented metrics that can be used for the assessment of various aspects of complex communications networks. This is a non-exhaustive summary of the existing metrics, and many more others are available in the literature. However, we consider this set of presented metrics to be the most fundamental and popular, enabling quick understanding of the rest, and in addition, it may constitute a strong basis for developing other more targeted metrics.

The centrality of a vertex may be determined according to any of the measures presented in the previous subsections of this chapter, based on the corresponding structural attributes of that vertex. The choice of a particular structural attribute and its associated measure depends upon the context of

Table 4.3: Complex and social analysis metrics at a glance.

Network	Degree distribution	Av Path Length	Centrality
Regular	dirac function	constant	constant
Small-world	varying	small	varying
Scale-free	power-law	usually small	varying
Random	Poisson	small	uniform
Random regular	uniform	long	uniform

the substantive application intended and it is usually a matter of conflicting constraints in each application framework of network science.

Chapter 5

Distinctive Structure and Features of Complex Networks

5.1 Network Structure and Evolution

The deep understanding of network structure and its evolution is of high research interest as it can be exploited in diverse applications, from generating realistic topological network models to designing efficient algorithms for the emerging underlying and/or overlaying networks. In this chapter, we describe and study analytically the most characteristic formations, properties and features that have been identified in complex networks along with various evolutionary processes characterizing and/or leading to them.

Also, while presenting the different social or other types of emerging structures and evolutionary processes, we will simultaneously provide useful insight of the importance and applicability of such features and properties in Network Science in general, and more targeted network types in particular. The term network structure captures all the properties of the network concerning the distribution of the degree of the nodes, the distance (in hops or other metrics) between node pairs, the connectivity, the density of the connections, the clustering coefficient, etc. The most characteristic structures of complex networks identified up until now, which are also examined in this book, are the following: the small-world phenomenon, the power-law degree distribution, the hyperbolic structure and the graph expansion. In addition, the models of their development, formation, and evolution, when applicable, are thoroughly examined.

The importance of understanding and categorizing the distinctive features of network structure and evolution processes lies in the immense opening potentials for designing precise and realistic network models and efficient algorithms for each category of network structure by exploiting such features. By understanding and controlling the (social or other distinctive) network structure and its evolutionary process, one can develop evolution models that lead to networks accurately approximating the real ones, such as the preferential attachment (Barabasi–Albert) model or the small-world model, which will be described and analyzed in Sections 5.2 and 5.3, respectively. In addition, the knowledge of node properties such as the degree distribution and clustering coefficient allows the localization of central or important nodes, while an estimation of the network average path length and the network expansion factor can provide significant insight on the efficient search potentials of the network or the delay of message passing. Moreover, connectivity features are exploited to spot bottleneck links or robust and vulnerable nodes and links. Last but not least, applications such as query passing, rumor spreading, node sampling, etc., may be designed and performed more efficiently with the aid of the knowledge of the structure and various properties of the network. The following sections will clarify the utility of developing characteristic network structure awareness by explicitly describing the latest observed structures and their impact on the network properties.

In this chapter, we focus on relational graphs, which are networks where any two nodes may possibly become neighbors, i.e., the connections between node pairs are not determined by their distances in a specifically defined metric space. Relational graphs represent systems of interactions, solely related to relation information, rather than defining such interactions through other factors as well (e.g. distance, weight, etc.). On the other hand, in the case of spatial graphs, the network is embedded in a metric space and the connected node pairs are selected according to their distances determined by the corresponding metric. An example of a relational graph is the social network of citations among researchers, while a wireless network constitutes a spatial graph due to the fact that nodes correspond to coordinates in a Euclidean space and their limited transmission power allows communications (connections) within particular distances and with certain constraints.

In the following sections of this chapter, we will describe and study in detail the most characteristic emerging complex network structures, along with their features, properties, and related applications.

5.2   Small-world Paradigm

5.2.1   Prolegomena—Description of a Small-World Network

In previous chapters, the notions of the average path length and the clustering coefficient are thoroughly explained (Section 4). In this section, we are going to study network graphs, which are distinctively characterized by high clustering coefficient and short average path length. A first question that arises regards the conditions/context needed so as to characterize the average path length value and similarly the clustering coefficient as low or high. Towards the search for measures of comparison for these two metrics, we revisit their definitions. Since the clustering coefficient is the probability of two connected nodes having a common neighbor, it can be compared with the probability of two randomly selected nodes in the network being connected, denoted as p. Therefore, the clustering coefficient can be characterized as high, if it is significantly higher than p. However, the probability p coincides with the density of the edges of the network (i.e., the number of edges existing in the network graph divided by the maximum possible number of edges, which is maximally equal to $\frac{n(n-1)}{2}$ for a network with n nodes). This in turn coincides with the connection probability and the clustering coefficient of a random graph (Sections 4.4, 2.2.1). As a result, the clustering coefficient of a network is high if it is significantly larger than the clustering coefficient of a corresponding random network with the same number of nodes and edges. Such a high value of clustering coefficient is as much as that of a regular lattice, and can be explained by the locality of the connections between node pairs. More precisely, in a regular lattice (see Figure 5.1(a)) each node connects only to close (local) neighbors in a specific, predetermined manner, which is the same for all nodes. This fact implies that it is very likely that the neighbors of a node are themselves neighbors leading to high values for the clustering coefficient (see Eq. (4.5) of Section 4.4). Therefore, with respect to the average path length, it is known that the average path length of a line graph (worst case scenario) grows linearly with the number of nodes n. Based on this observation, we can consider that a type of network graph has short average path length, if its average path length increases slower than linearly with respect to n, i.e., as $\log(n)$ [55, 102]. As is already known, random graphs are characterized by average path length that grows logarithmically with n [72], and due to this fact, they can be considered as a point of reference for the comparisons regarding the network path length. More precisely, if a network graph has an average path length comparable with that of a random graph with the same number of nodes and edges, then it is considered as having a short average path length.

The term "small-world" is used to characterize networks that are neither regular nor random. They can be highly clustered, like regular lattices, but at the same time have small characteristic path lengths, like random graphs. They lie somewhere between these two extremes. From a mathematical point of view, a small-world network refers to a graph growing in size, the average path length of which increases proportionally to the logarithm of the number of nodes in the network. From a practical point of view, in the context of an online social network (Internet, citation network), the small-world phenomenon corresponds to strangers being linked through a small number of individuals. In the following, we will describe two famous experiments of decentralized search in real networks, both of which gave rise to the extensive research of the small-world phenomenon in the network structure, and in the sequel we will thoroughly study the two small-world models that have prevailed in the literature.

5.2.2   Large-scale Experiments—"Six Degrees of Separation"

The term "six degrees of separation" is due to the experiment of Travers and Milgram [154] in 1969, which was the first attempt to uncover the interconnections and characterize the navigational properties of a large social network such as American society (about 200 million residents at the time of the study). Travers and Milgram performed a chain-letter experiment, aiming to examine the probability that two strangers in the United States are connected via a short path length and how characteristics such as geographic location and occupation of the two strangers affect the composition of the individuals lying on the path. In the procedure they designed, they selected three groups of starting people and a target person, and a document was given to each starter along with the requirement to forward it towards the target. The target person was a stockbroker in a suburb of Boston, Massachusetts. Two starter groups were chosen from Nebraska; one random group of 96 volunteers and one group of 100 stockbrokers (the same occupation (affiliation) circle as the target). Also, one more random starting group was chosen from the same geographic area as the target, i.e., Boston. The diffusion of the packets took place as follows. In the case that the starter knew the target personally (they have met before and knew each other on a first name basis), the starter should have mailed the message directly to the target. Otherwise, the starter should have sent it to an acquaintance (friend or relative) satisfying two requirements: (a) the sender should know the acquaintance on a first-name basis, (b) the acquaintance chosen is perceived as more likely to know the target or forward the document closer to the target (due to different criteria such as geographic distance, similar occupation, etc.). For all the messages that reached the target, the researchers could count the number of times that they were forwarded and could also collect the names and other characteristics (such as age, occupation, sex, etc.) of people lying on the intermediate path, i.e., reconstruct the whole human path that took place.

From 296 letters sent, 64 letters eventually reached the target in Boston. The average path length was around six (in fact equal to 5.2), which can be considered as small relatively to the population of the United States. Hence, the researchers concluded that people in the United States are separated by short paths of about five intermediates on average (distance of six hops). According to the researchers' evidence, the dropout of the messages can be characterized as "random" (regarding the information about the participants such as age, sex, etc.) and are either due to a participant's apathy or unwillingness to forward the document in a following step or due to his/her difficulty in finding a suitable next hop. In addition, as expected, the completed chains started from Boston were shorter on average than those started in Nebraska, and similarly the completed chains initiated by stockbrokers in Nebraska were shorter than those initiated in the same geographic area by random individuals. These results indicate how social and geographic location parameters can influence the length of the paths separating two US residents. This experiment, except for showing the small average path length separating a pair of strangers in a large community, was also one of the first real experiments of decentralized search in networks proving that people can find the existing short paths without full knowledge of the network. Indeed, the existing short paths in a social network are out of interest (i.e. not useful) if they cannot be discovered in a decentralized way, as individuals in a social network have only local information about the global network [52]. The basic conclusions of Travers and Milgram's experiment—the existence of short paths connecting two strangers and the ability to locate these paths in a decentralized way—are reflected in the two basic models for small-world graphs that are examined in the following subsections (Watts and Strogatz model in Section 5.2.3 and Kleinberg's model in Section 5.2.4). Finally, it was observed that the penultimate individuals of the 64 completed chains were actually 26 people, in the sense that the same acquaintances of the final target forwarded directly to him multiple documents from different senders. This is evidence of the overlapping of the various paths and the existence of a few node-hubs with high degree (high circle of acquaintances) able to contract different paths. This conclusion can be jointly considered with the emergence of heavy-tailed distributions in social networks [12], i.e., the fact that there are a few nodes with high degree and many nodes with low or medium-range degree (as will be discussed in Section 5.3).

More recently (2003) a similar but more extensive experimental study was performed by Dodds, Muhamad, and Watts [52], who constructed a global Internet-based social search experiment. The experiment described and analyzed in [52], considered 18 targets instead of a single target of the Travers and Milgram experiment. The participants registered online for the experiment and they were assigned a target and asked to forward the message towards the destination via email. Although the participants created about 24, 000 distinct message chains, which is a much larger volume that the hundreds of chains initiated by Travers and Milgram's experiment, the completion rate of the experiment did not increase, since only 384 chains reached their corresponding destination nodes. This study ([52]) re-discovered many of the conclusions of [154], such as the small average path length of the network and the random failure of the chains. However, there were no indications in the results about node-hubs, each one of them constituting the penultimate hop for many message transfers to the destination. The main conclusion was that in order for the small-world hypothesis to hold, except from the existence of short paths separating individuals, the members of a social network should have enough incentives to forward the messages and allow for a decentralized search with short paths. This was verified by increasing the incentives of the participants and observing the subsequent decrease in the length of the paths.

In the following subsections, we are going to describe, analyze, and discuss particular mathematical models for the formation of network graphs with small-world properties. Further subsections will be devoted to providing quantitative means for studying and exploiting the small-world phenomenon in complex networks analysis/control and more specifically in wireless complex communications networks.

5.2.3   Watts and Strogatz Model (WS Model)

The Watts and Strogatz model [162], is a constructive process for obtaining small-world and some other types of random graphs, beginning with a regular lattice structure. This constructive model led to a formal definition of small-world networks, as was mentioned before (Section 5.2.1), which will be more precisely formulated in the sequel. To provide some insight, the model of [162] starts from a clustered structure (regular lattice) and adds random edges connecting nodes that are otherwise far apart in terms of hop distance. These random long edges will be denoted as "shortcuts" for the rest of the chapter. The initial clustered structure ensures a high clustering coefficient for the final network, while a suitable number of added shortcuts can further reduce the average path length, up to a sufficient level, so that the created graph may be characterized as small-world. The author in [162], presents three models, denoted by a-model, β-model, and ϕ-model, each of them characterized by a single parameter a, β, ϕ respectively. The value of each parameter in each model defines a set of topologically similar graphs in the sense that their average path length and clustering coefficient depend only on the number of nodes n and the average node degree k of the graph. Each one of these one-parameter models leads to a family of graphs with topologies varying from totally ordered structure to complete randomness. Therefore, these models provide the flexibility to dictate the properties of the final topology (through the corresponding parameter) from the regular structure to the small-world structure and then to random networks, and to study interesting emerging scaling properties and phase transitions (or threshold phenomena) regarding the average path length and the clustering coefficient of the yielded networks. In the following, we are going to describe and study the simplest of these models, namely the β-model, also denoted by Watts and Strogatz model [163], [162].

The β-model's algorithm modifies the topological structure of an ordered lattice by randomly rewiring edges with increasing probability β up to the point that a random graph's topology is achieved. Specifically, the β-model starts with the totally ordered structure of a ring lattice, where each node i has k neighbors, the set $\{i-\frac{k}{2}, ..., i-1, i+1, ...i+\frac{k}{2}\}$, where we use the symbol $i-h$ for a node being h positions far away from i in the anti-clockwise sense and similarly for $i+h$ in the clockwise sense (see Figure 5.1(a)). The algorithm takes $\frac{k}{2}$ rounds, each consisting of n sub-rounds. In a round j, for each of its sub-rounds, a vertex i of the graph is selected in turn along with the edge connecting it to its $i+j$ neighbor (clockwise sense). With probability β the edge $(i, i+j)$ is rewired from node $i+j$ to a randomly and uniformly selected node of the lattice. The n sub-rounds span all the vertices of the graph exactly once per vertex, where vertices are selected in turn in a clockwise sense. After all rounds have finished, all the edges of the graph have been traversed exactly one time each. The process is shown in Figure 5.1. In case $\beta = 0$ (Figure 5.1(a)) the graph remains a 1-lattice (ring with $k = 4$), while, the case of $\beta = 1$ modifies the initial ring to a completely random graph (Figure 5.1(c)). The cases $0 < \beta < 1$ (Figure 5.1(b)) are of particular interest, as they can lead to network graphs between the two extremes of completely ordered and random graphs (with respect to the values of the average path length and the clustering coefficient). Parameter β controls the degree of randomness of the graph [162], since it dictates the transformation of the initial regular lattice to a random graph with asymptotically known properties.

In Figure 5.2, the scaling behavior of the average path length and the clustering coefficient with respect to the parameter β (beta) are presented. It can be observed that the average path length $L(\beta)$ (semilog scale) is characterized by a rapid transition between its corresponding values for the two extreme cases of the ring lattice and a random graph, and this transition corresponds to a relatively small value of β. A similar tipping phenomenon characterizes the clustering coefficient $C(\beta)$, but for a higher value of β, which means that the clustering coefficient remains high for a long interval of parameter's β values, after $L(\beta)$ has approached its value for random graphs. This difference in transitions happens due to the fact that the rewiring of an edge creates a new edge that serves as a shortcut for two previously long-distance nodes and their neighborhoods. In a large graph, only a few shortcuts can contract the distances between widely separated parts leading to a sudden, high reduction of the average path length. However, the clustered character of the initial ring, as described by the value of the clustering coefficient, does not change rapidly with the addition of a small number of shortcuts, since a few edges removed from their corresponding triads do not provoke sudden decrease in the clustering coefficient (see Eq. (4.5), Chapter 4). This observation gave rise to the abstract definition of the small-world graph, as follows:

Definition 18 *(Small-world graphs) There exists a class of graphs that are highly clustered, yet have characteristic path length and length scaling properties equivalent to random graphs. These are called small-world graphs ([162]).*

In other words, for the intermediate values of β between the transition of the average path length and the transition of the clustering coefficient to their random limits (i.e., their values for random graphs), small-world phenomenon properties emerge. Each value of the parameter β corresponds to a different type of graph structure, from totally ordered graphs to small-world graphs and to totally random graphs leading to a graph structure continuum from regularity to randomness. By choosing the desirable value of β we can produce a specific type of graph from the whole continuum or family of network structures. If instead of rewiring links to create shortcuts, we add new links

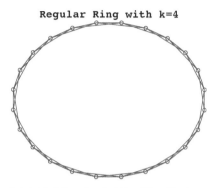

(a) Initial Regular Lattice ($\beta = 0$), Average path length = 3.4 hops, Average clustering coefficient = 0.5.

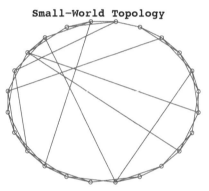

(b) Small-World Graph ($\beta = 0.2$), Average path length = 2.5 hops, Average clustering coefficient = 0.3.

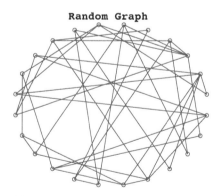

(c) Random Graph ($\beta = 1$), Average path length = 2.5 hops, Average clustering coefficient = 0.01.

Figure 5.1: Demonstration of the β-model of Watts and Strogatz. The network parameters are $n = 24$, $k = 4$.

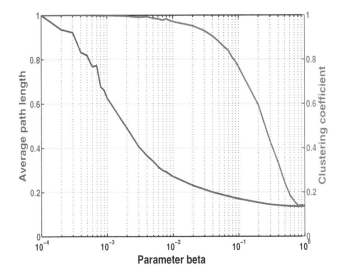

Figure 5.2: The β-model of Watts and Strogatz: Average path length and clustering coefficient scaling with varying β (beta) parameter. The network parameters are $n = 1000$, $k = 10$.

that function as shortcuts, we conclude similar results. According to [162], a more specific definition of the small-world graph can be given only in terms of its corresponding random graph (i.e., with the same number of nodes and expected number of edges) as follows:

Definition 19 *(Small-world graphs) A small-world graph is a graph with n-vertices and average degree k that exhibits $L \approx L_{random}(n,k)$, but $\gamma \gg \gamma_{random} \approx \frac{k}{n}$, where $L_{random}(n,k)$, $\gamma_{random} \approx \frac{k}{n}$ are the average path length and the clustering coefficient of the random graph with n-vertices and average degree k correspondingly ([162]).*

5.2.4 Kleinberg's Model

In contrast to the Watts and Strogatz model, Kleinberg's model [99] consists of two parts: a constructive one, similar to the Watts and Strogatz constructive process, and an algorithmic one. Kleinberg observed that Milgram's experiment made in fact two discoveries. The first was the discovery of short paths separating communicating node pairs, while the second was the ability of nodes (participants in the experiment) to find these short paths by using local only information. Graphs with both these properties are denoted as small-world navigable graphs.

With respect to the Kleinberg model for small-world navigable graphs, the starting network topology design resembles in concept the corresponding one of Watts and Strogatz. The main difference of Watts and Strogatz and Kleinberg's topology is that the former starts from a regular ring where some edges have been rewired to form long range links, while the latter starts with a two dimensional grid enhanced with additional long range links. More precisely, the network is a square 2-dimensional grid topology with n^2 nodes, enhanced with the lattice distance, i.e., if i, j are two nodes with coordinates in the grid (x_i, y_i) and (x_j, y_j) correspondingly and $d(i, j)$ is their lattice distance then $d(i, j) = |x_i - x_j| + |y_i - y_j|$. Each node is connected with all its neighbors in a specified lattice distance p, forming in this way local connections that increase the network clustering coefficient (Figure 5.3). Simultaneously, each node forms directed long-range connections or shortcuts with q other nodes (Figure 5.3). The probability that the ordered edge (i, j) is formed is proportional to $|d(i, j)|^{-r}$ where the control parameter r determines how far one can move in the network through a shortcut. As a result, if $r = 0$, each node is connected with q other nodes selected with equal probability among all nodes (uniform distribution of long-range edges). In this case, the shortcuts are very likely to be long, thus further contracting the network paths. As r increases, the shortcuts tend to be shorter, providing a much smaller advancement in reaching a node located far away. Therefore, as also stated in [99], if the parameters p, q are considered stable, the Kleinberg model is a mono-parametric model—function of the parameter r—which leads to a family of graphs, where each member is determined by the value of the r parameter.

Although Kleinberg's constructive model resembles the small-world network construction of Watts and Strogatz, Kleinberg proved that only a unique member of this family of graphs is small-world navigable, i.e., graphs where the nodes can potentially find the existing short paths using only local information. Local information is meant in the sense that the message holder at each step of the algorithm's deployment knows only its local connections, the position of the destination node in the grid and the locations and long-range contacts of all the nodes that have pre-occupied the message. As we know, the average path length in a small-world graph is polynomial in $\log n$. However, Kleinberg proved that only if $r = 2$ in the predefined family of graphs, the expected delivery time of a decentralized algorithm is at most $a_0(\log n)^2$, i.e., polynomially dependent on $\log n$ similarly to the average shortest path length, where the expected delivery time of an algorithm is the average number of steps taken by the algorithm to deliver the message from source to destination. Otherwise, if $r < 2$, the expected delivery time of a decentralized algorithm is at least $a_r n^{2/3}$ and if $r > 2$, the expected delivery time of a decentralized algorithm is at least of order $a_r n^{(r-2)/(r-1)}$, where the constants a_0, a_r are independent of n. Also, in the case of $r = 2$, the decentralized algorithm that achieves the $a_0(\log n)^2$ upper bound, routes a message to the one-hop neighbor closer to the destination in lattice distance, i.e., it is a simple greedy algorithm. Therefore, by exploiting additional information, breaking slightly the restriction for local information, one can achieve a lower expected deliv-

Kleinberg's Model for Small—World Graphs

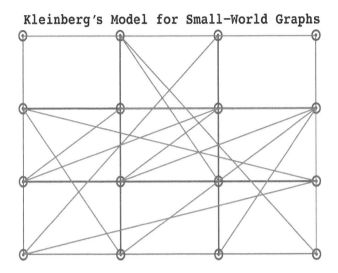

Figure 5.3: Small-world navigable graph of Kleinberg's Model. The topology is a 2-regular grid consisting of $n = 16$ nodes with $p = 1$ and one directed long range edge beginning from each node ($q = 1$). For some nodes the corresponding shortcut is not obvious due to being a vertical or horizontal shortcut coinciding with already existing grid connections.

ery time at a relevant signaling cost paid for obtaining that extra non-local information.

5.2.5   Examples and Applications

For plenty of networks appearing in our daily life, in nature, in science, etc., the characterization "small-world network" or "small-world navigable network" matches exactly with their structure. In this subsection, we refer to relevant examples and describe some of their prominent small-world properties. A synopsis of some of these examples is shown in Table 5.1. To provide an intuition about a real life small-world network, let us imagine a road map of a country or a big city. One can observe a clustered structure consisting of small roads of different areas enhanced with highways, which connect areas otherwise separated by a large number of small roads, functioning as shortcuts. A road map can be represented by a network graph where the nodes are the roads and two roads are linked if they cross each other. This is a highly clustered network, with small average path length due to the existing shortcuts or highways (which correspond to nodes with high degree in the aforementioned graph). Indeed, a roadmap network is a small-world one that we encounter in our everyday transportation and it resembles the small-world navigable graph constructed by Kleinberg (Section 5.2.4). Following the same idea, we can also assign the characterization small-world to the social influence graphs, i.e., graphs where the nodes correspond to individuals and a directed link from a node to another one is added, if the second is influenced by the first. One's ideas and thoughts are mostly influenced by one's close friends or relatives, but also by commonly known people such as politicians, artists, etc., who are well-connected individuals that "shorten" the distances of social influence. In the class of engineered networks, a heterogeneous wireless/wired ad hoc network consists of both wireless parts and wire-based connections that connect long-distance wireless nodes. This combination also yields a small-world structure, since the wireless ad hoc networks are already highly clustered due to the locality of their connections, and in addition, the long wires connect directly some previously long-distance nodes, otherwise separated by many wireless hops.

Table 5.1 presents some examples of networks that have been identified as "small-world" through experimentation and statistical analysis [6, 115]. For each one of them, the table denotes their type and description in the first two columns and the absolute value of the average path length L_{avg} on the third column. The fraction $\frac{L_{avg}}{L_r}$, where L_r is the average path length of a random graph with the same number of nodes and links, compares the average path length of the network L_{avg} with that of the corresponding random graph and indicates a small-world structure if $\frac{L_{avg}}{L_r}$ is close to 1. Similarly, the fraction $\frac{C_{avg}}{C_r}$, where C_r is the clustering coefficient of the corresponding random graph and C_{avg} the clustering coefficient of the network under analysis, indicates a small-world structure if $\frac{C_{avg}}{C_r} >> 1$. Finally, we provide the size of each network for allowing comparisons with its absolute average path length. All the networks exhibit emerging small-world properties as can be deduced by computing the quantities $\frac{L_{avg}}{L_r}$, $\frac{C_{avg}}{C_r}$. Also their average path length is disproportional to their size. The first network consists of movie actors as nodes, who are linked through an edge if they have acted in the same movie together [163]. The second is the network of the World Wide Web [2], where the nodes represent domain names and two nodes are connected if any of the Web pages in one domain is linked to any Web page in the other. The three following networks correspond to data from networks of scientists [14], where nodes are scientists and there is an edge between two scientists if they have coauthored a paper. The fourth and the penultimate network both refer to collaboration of mathematicians. However, the fourth is constructed as in [14], while the fifth as in [131]. Finally, the power grid network of the western United States [163] consists of nodes that model generators, transformers, and substations, and edges modeling high-voltage transmission lines.

Table 5.1: Examples of small-world networks. L_{avg} is their average path length and C_{avg} their clustering coefficient, while L_r and C_r are the corresponding metrics for a random graph with the same number of nodes and edges (expected number of edges). The parameter size refers to the largest connected component of the network.

Network	Description	L_{avg}	L_{avg}/L_r	C_{avg}/C_r	Size (giant component)	Reference
Movie Actors	*Nodes*: Actors; *Edges*: Link actors that have acted in the same movie	3.65	1.22	2,925.93	225,226	[163], [6]
WWW (undirected)	*Nodes*: Separate Domain Names; *Edges*: Link two nodes if any of the pages in one domain is linked to any page in the other	3.1	0.93	4,468.7	153,127	[2], [6]
Neuroscience coauthorship	*Nodes*: Neuroscientists; *Edges*: Link Neuroscientists if they own a joint publication	6	1.2	13,818.18	209,293	[14], [6]
Math coauthorship	*Nodes*: Mathematicians; *Edges*: Link Mathematicians if they have a joint publication	9.5	1.15	10,925.93	70,975	[14], [6]
Erdős Numbers (similar to math coauthorship)	*Nodes*: Mathematicians; *Edges*: Link Mathematicians if they have a joint publication	7.64	1.26	4,416.4	268,000	[131]
Power grid	*Nodes*: Generators, transformers, substations; *Edges*: High-voltage transmission lines	18.7	1.5	16	4,941	[163], [6]

5.3   Scale-free Networks

5.3.1   Definition and Properties

In this section, we will study another type of complex network bearing social features, closely related to the small-world structure (Section 5.2), which can also be used to reduce the average path length of other complex networks. The "scale-free" property of a class of complex networks is closely related to the notion of popularity [57] and it is also tightly related to the degree distribution of several social networks. Specifically, the concept of "scale-free" captures the lack of scale in the degree distribution of complex networks. More precisely, in scale-free networks, different node groups exhibit differences in scaling of their node degree, interpreted as scale difference in connectivity and neighborhood relations.

To begin with, until recently, all complex networks were thought to be completely random (i.e., Erdős and Rényi type Random Graphs). In random networks, the node degree distribution follows a Poisson distribution with a bell shape as shown in Figure 5.4. According to the Poisson distribution, the probability of a node to have k connections decreases exponentially for large k. Thus, it is extremely rare to find nodes having significantly more or fewer links than the average. Approximating the degree distribution of a complex network with the normal distribution is closely related to the Central Limit Theorem, according to which the limit of the sum of small independent, egalitarian, random quantities follows the Normal distribution [127]. This would also be the case for complex networks subject to two important assumptions. The first one [57] is that the formation of a link between two nodes should be independent of the connection of every other node pair, so that we ensure the independence of connections. As a counter example for this assumption (not the only one), let us consider the small-world networks (Section 5.2), which present high clustering coefficients, i.e., meaning that the probability of two nodes being connected depends on their already existing mutual connections. Therefore, a frequently made assumption, namely that of edge independence in the network formation phase, does not hold for all complex networks. Secondly, the degree of a node should be considered as the sum of these independently formed links [57], which in this case does not hold, due to the circumvention of the independence of link formation.

However, complex networks display more complicated features regarding their structure and architecture. Their complexity lies mainly in the way nodes are inter-connected as well as the interactions among them as the network evolves. More specifically, contrary to the assumptions of the Central Limit Theory, the behavior of the nodes regarding the formation of new connections, or modification of existing ones, is correlated across the population and as a result, the node degrees are characterized by imbalances. As an example, in human communities, there are usually a few people who are very popular and well-known to most of the world, such as politicians, actors, etc., and many more others (the majority) known to only to a very close circle, e.g., family, friends and work circles.

In general, complex networks evolve according to two basic mechanisms that are not taken into consideration by the random graph model. The first one is "growth." The way a network evolves, indicates that new nodes tend to link to existing ones. The second is what is most popularly known as preferential attachment, the fact that when nodes form new connections, they tend to connect to other nodes with probability proportional to the popularity of the existing ones. Since not all nodes are equally popular, some of them are more desirable than others. More explicitly, the probability that a new node connects to existing ones is not uniform, but instead favors the nodes displaying larger connectivity or degree, leading to a degree-heterogeneous system with two extreme groups of hub nodes and low-degree nodes as shown in Figure 5.6. The powerful, high degree nodes, referred to as hubs, have a seemingly unlimited number of links, while the vast majority of nodes have a small number of links (Fig 5.5). Preferential attachment is usually considered to be linear with respect to node degree; however, other types of preferential attachment such as non-linear or preferential attachment combined with initial fitness have also been examined [6]. These cases will be discussed in the following sections.

It is derived that large networks self-organize into a scale-free state. This is due to both the two mechanisms of growth and preferential attachment, as will be shown in section 5.3.3. The probability that a vertex interacts with k other vertices is a power-law, following a model such as $P(k) \sim k^{\gamma}$, with an exponent γ between 2.1 and 4, as shown through experimentations in real world networks [6]. Power-law distributions are quite different from the bell-shaped distributions that characterize random networks. A power-law function is continuously decreasing and does not have a peak, contrary to a bell shaped curve. Consequently, the distribution of links is not democratic in contrast to random networks. Indeed, there are a few hubs that dominate because of having a large node degree, while there are many nodes with a small degree. The random and power-law degree distributions are compared in Figures 5.4 and 5.5. Each one of them is shown in both linear–linear scale and logarithmic–logarithmic (log–log) scale. For the Poisson degree distribution, Figure 5.4(a) depicts the distribution in linear scale in both x-axis and y-axis, while Figure 5.4(b) illustrates the distribution in logarithmic scale. Correspondingly, Figures 5.5(a) and 5.5(b) show the curves in the two scales for the power-law distribution. The reason for including the log-log scale in the presentation is that it is important to emphasize the linearity of the power-law distribution curve in this case, in contrast to the Poisson distribution. This linearity is often called the "signature" of the scale-free networks and it is used to verify the scale-free structure of real large scale networks under study through collected data sets, by plotting their degree distributions in log–log scale.

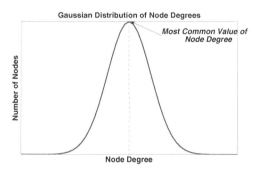

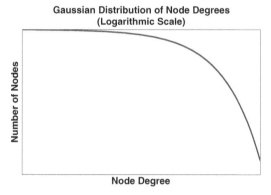

(a) Poisson degree distribution. There is a strict cut-off up to a specific value of node degree.

(b) Poisson degree distribution in logarithmic scale.

Figure 5.4: Random graph degree distribution.

We note that scale-free networks reveal two significant properties: they are remarkably resistant to accidental failures, but extremely vulnerable to coordinated attacks [13]. The first one happens due to the fact that the majority of nodes in a scale-free topology have a small number of connections. Therefore, a failure to a random node would damage, with high probability, a node with a small degree and therefore a node whose importance in the network is restricted. However, if one wants to attack a scale-free topology, he/she will address his/her attack to a hub, so as to have higher probability to damage the network, as hubs play an important role in maintaining network connectivity. It is important to notice that the hubs can potentially decrease the hop-distance by linking the neighborhoods of otherwise long-distance nodes, especially in highly clustered networks (which refers to the notion of contraction as this is described in [163]). In this way, they give rise to the small-world phenomenon, by reducing distances in a clustered structure.

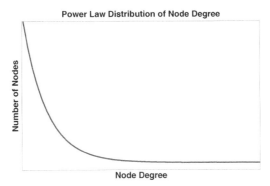

(a) Power-law degree distribution. Nodes with high degree exist with a small probability.

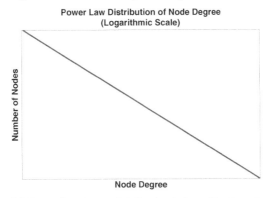

(b) Power-law degree distribution in logarithmic scale.

Figure 5.5: Degree distribution of scale-free types of complex networks.

5.3.2 Examples and Applications

As in the case of small-world networks in Section 5.2.5, we will describe some examples of scale-free networks emerging in real situations, in order to make the definition and properties of the scale-free paradigm more tangible. Table 5.2 presents selected features and properties of some real scale-free networks, as these features were obtained through statistical analysis of collected datasets. The coauthorship networks (math and neuroscience) and the movie actors' network adopt the same description as in the examples for small-world networks (Section 5.2.5). However, the topology of the World Wide Web in [7] is examined through a different point of view than in Section 5.2.5. The nodes represent Web pages and the edges hyperlinks, which appear when a Web page points to another. Finally, the citation network consists of nodes representing published articles and links that point from an article to another if the first has a reference to the second. If the network is directed (such as the

WWW and the citation network), for each node two degrees are defined, its in-degree and its out-degree and therefore there are two degree distributions (in and out) and two computed power-law exponents; γ_{in} for the in-degree distribution and γ_{out} for the out-degree distribution. In undirected networks the in and out degree distributions coincide and therefore the power-law coefficients coincide $\gamma_{in} = \gamma_{out}$ as well. As it can be observed in Table 5.2, all the networks under examination have power-law degree distributions, as their estimated power-law exponents are different than 1 and for the first five networks, which are social networks, the power-law exponents vary between 2 and 3. For instance, the air transportation network is an artificial network, however it is still characterized by an exponent other than unity.

An interesting scale-free and small-world network is the worldwide air transportation network [79]. The interest in this network is due to its major role in the propagation of commodities, diseases, the mobility of millions of passengers every day, the popularity for a country based on its airport sizes and importance, etc. The nodes of the airport graph represent the cities (we consider that each node covers all the airports of each city) and there is an edge connecting a city to another one if there is a direct flight from the first to the second. The degree of a city is the number of other cities for which it has a direct fight (like the out-degree of a directed network). More specifically, on the one hand, there are small airports with, e.g., only domestic flights inside the corresponding country, having therefore low connectivity in the airport graph. On the other hand, there are airports-hubs, especially in major capitals or large cities, operating both local and international flights and linking long-distance countries, in which case the links are obviously functioning as shortcuts. Indicatively, the average path length of the air transportation network, the details of which are presented in Table 5.2, is only 4.4, which is very small compared to the size of the graph, and it is found to grow logarithmically with the number of the cities. Its clustering coefficient is 0.62, while for the corresponding random graph as defined in Section 5.2.5, it is much smaller and equal to 0.049. Therefore, the air transportation graph is characterized by both small-world and scale-free properties (Table 5.2).

5.3.3   Barabási–Albert Model

During the past few years, significant interest has been observed for developing models that mimic the evolution dynamics (node/edge addition/deletion) of real complex networks and construct network topologies with complex networks' characteristics such as scale-free or small-world properties. In contrast to the Watts and Strogatz model, these models do not enhance an already existing topology, but they construct a topology from scratch. A category of these models, called "Preferential Attachment Models," uses preferential attachment laws to create networks with power-law degree distributions. In this section, we provide a heuristic version of the "Preferential Attachment

Table 5.2: Examples of scale-free networks. We denote with $<k>$ the average node degree, with γ_{out}, γ_{in} the exponents for the out and in-node degrees correspondingly, and the parameter size refers to the largest connected component of the network.

Network	Description	$<k>$	γ_{out}	γ_{in}	Size (giant component)	Reference
Movie Actors	*Nodes*: Actors; *Edges*: Link actors that have appeared in the same movie	28.78	2.3	2.3	212,250	[1], [6]
WWW (directed)	*Nodes*: Web pages; *Edges* (*directed*): Point from a Web page to another if the first has a hyperlink to the second	4.51	2.45	2.1	325,729	[7], [6]
Neuroscience coauthorship	*Nodes*: Neuroscientists; *Edges*: Link Neuroscientists if they own a joint publication	11.54	2.1	2.1	209,293	[14], [6]
Math coauthorship	*Nodes*: Mathematicians; *Edges*: Link Mathematicians if they have a joint publication	3.9	2.5	2.5	70,975	[14], [6]
Citation network	*Nodes*: Published articles; *Edges*: Represent references to previously published articles	8.57	–	3	783,339	[134], [6]
Airport network	*Nodes*: Cities; *Edges*: Link two cities if they are connected by a direct flight	–	1.0 ± 0.1	–	3,663	[79]

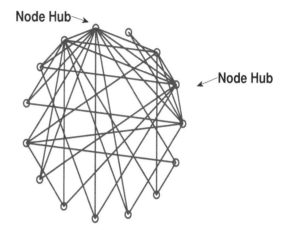

Scale–free network with 15 nodes (BA Model).

Figure 5.6: Visualization of a scale-free graph created by the Barabási–Albert Model with 15 nodes and 3 connections per new node.

Models" that uses the mean field theory from physics to show that the combination of growth and preferential attachment leads to a power-law degree distribution. The following model is denoted as Barabási–Albert Model (BA Model) and by being the first evolutionary model using preferential attachment that emerged, it is probably the most representative of its category. The BA model, instead of transforming a network topology to enhance it with particular features, as in the Watts and Strogatz model, is focused on modeling the evolution dynamics of the network, and the final topology is obtained as an outcome of this evolution.

Barabási and Albert incorporated in their evolutionary model two basic mechanisms, growth and preferential attachment, as mentioned in Section 5.3.1. In this way, the obtained model dynamics essentially captured the dynamics of the real-world networks. In general, through observations and statistical analysis, real world networks have been observed to grow in size with time, while the probability to connect to a specific node depends on the node's popularity (degree). Examples of real-world networks characterized by such evolutionary dynamics are shown in Table 5.2, i.e., the World Wide Web or the Citations' network. In the two following subsections we will elaborate more on the two basic mechanisms of the BA's model, growth and preferential attachment. These two basic ingredients of the BA's model are shown through continuum theory to lead to scale-free or power-law degree distributions.

Preferential Attachment

Suppose that each node i of the network is characterized by a parameter x_i with a specified range of values. The parameter x_i usually represents the node degree, but for example it may also be the global trust value of the node or it may also represent another characteristic of a node. We define as linear preferential attachment with respect to the parameter x_i, the process of selecting a node with probability proportional to the value of the parameter x_i, expressed as follows:

$$\Pi(x_i) = \frac{x_i}{\sum_{\forall j} x_j} \tag{5.1}$$

The denominator sum is a normalization factor in order to obtain a probability distribution across all nodes in the network. In Barabási–Albert model, the parameter x_i for each node i coincides with its degree denoted as k_i, i.e., $x_i = k_i$.

In the general case, the preferential attachment rule can be expressed in terms of functionals depending on each x_i, according to the following formulation:

$$\Pi(x_i) = \frac{f(x_i)}{\sum_{\forall j} f(x_j)}, \tag{5.2}$$

where $f(x_i)$ is an increasing function of the quantity x_i. We will discuss more about this formulation in Section 5.3.4.

Growth

In most cases, real networks are observed to grow in size with time, e.g., communications networks, social networks, etc. More specifically, with the term "growth" we refer to the addition of new nodes in the network, which connect to already existing nodes. This process will be denoted from now on as node addition. In the BA model, it is considered that an initial core network exists and during the network evolution, new members join the graph by forming links with existing members, under the rule of linear preferential attachment. The process of growth will become clearer in the following, as we describe in more detail the BA's algorithm.

Barabási–Albert Algorithm

At this point we will describe the steps followed by the BA's model [6]. The time is considered slotted. The starting network consists of m_0 nodes. Each newcomer connects to the network through m connections with m appropriately chosen nodes (one for each chosen node), already present in the network. Thus, at each time slot t:

- A new node joins the network and connects to m existing nodes.

- Each of the m existing nodes is chosen according to the preferential attachment rule, i.e., the node i with degree $k_i(t)$ is chosen with probability $\Pi(k_i) = \frac{k_i(t)}{\sum_{\forall j} k_j(t)}$, where the sum at the denominator spans all network nodes.

- The number of nodes in the network is equal to $N_t = m_0 + t$.

A scale-free graph, created by the BA's algorithm, is depicted in Figure 5.6. Considering the average path length and the clustering coefficient of a scale-free graph, they are both shown through simulations [6] to be significantly less and higher, respectively, than those of a corresponding random graph (as defined in Section 5.2.5). Due to these observations, one may reach the conclusion that the scale-free graphs present a small-world behavior, which however, is not always the case. Although their average path length increases logarithmically with the number of nodes, the clustering coefficient decreases with the number of nodes as well instead of being fixed [6].

Continuum Theory

In this subsection, we provide the derivation of the mathematical expression of the average degree k_i of each node i, in the network produced by the Barabási–Albert Algorithm. We employ the continuum theory approach [6] [126], which considers k_i as a continuous variable and approximates its variation with time t (time dependence), with a differential equation. According to the BA model, only the processes of node addition and link addition are permitted, meaning that the network only increases (each new node also adds $2m$ directed links), therefore the average degree increases with time. Through the continuum theory approach [126], we prove the power-law relation of k_i with time t. We assume that the node i joins the network at time t_i.

For all i, the average degree k_i satisfies at each time slot t the differential equation:

$$\frac{dk_i(t)}{dt} = m\Pi(k_i) = m\frac{k_i(t)}{\sum_{j=1}^{N_t-1} k_j(t)} \tag{5.3}$$

where the sum in the denominator excludes the node added at time slot t, while also $N_t = t$ leading to $\sum_{j=1}^{N_t-1} k_j(t) = 2mt - 2m$. Solving Eq. (5.3) with the initial condition $k_i(t_i) = m$, gives

$$k_i(t) = m\left(\frac{t}{t_i}\right)^{\frac{1}{2}} \tag{5.4}$$

which indicates that $\forall\ i$, k_i changes according to a power-law of t with exponent $\frac{1}{2}$.

In the sequel, we deduce the power-law probability distribution (pdf) of the average node degree $k_i(t)$, $P(k)$ (Appendix B). The cumulative distribution of $k_i(t)$ (cdf) is:

$$P(k_i(t) < k) = P\left(m\left(\frac{t}{t_i}\right)^{\frac{1}{2}} < k\right) = P\left(t_i > \frac{m^2 t}{k^2}\right) \tag{5.5}$$

Assuming that the nodes enter the network at regularly spaced time internals, $P(t_i) = \frac{1}{m_0 + t}$. Therefore,

$$P\left(t_i > \frac{m^2 t}{k^2}\right) = 1 - \frac{m^2 t}{k^2(t + m_0)} \tag{5.6}$$

As a result, the pdf can be expressed as:

$$
\begin{aligned}
P(k) \quad &= \quad \frac{dP[k_i(t) < k]}{dk} \\
&= \quad \frac{2m^2 t}{k^3(t + m_0)} \\
\xrightarrow{t \to \infty} \quad &\quad 2m^2 k^{-3}
\end{aligned}
\tag{5.7}
$$

which corresponds to a time-independent power-law degree distribution with an exponent equal to -3, leading to the conclusion that the network produced by the BA's model reaches a stationary scale-free state.

Apart from continuum theory, there exist other methodologies for proving the power-law degree distribution. One of them is the "Rate equation" approach [6, 100], which, rather than focusing on the average degree k_i, approximates the rate of change of the number of nodes with degree k at time t, $N_{k,t}$, with a differential equation. It uses the observation that any increases in $N_{k,t}$ can happen when nodes with degree $k - 1$ obtain a new edge, while any decreases in $N_{k,t}$ happen when nodes with degree k obtain new connections. Therefore, the cumulative rate equation is:

$$\frac{dN_{k,t}}{dt} = m\frac{(k-1)N_{k-1,t} - kN_{k,t}}{\sum_{\forall j} jN_{j,t}} + \delta_{k,m} \tag{5.8}$$

The Kronecker $\delta_{k,m}$ corresponds to new nodes added with degree m and it holds that $\delta_{k,m} = 1$ if $k = m$, while if $k \neq m$, $\delta_{k,m} = 0$. The first summand of the right hand side of Eq. (5.8) corresponds to the observation stated just above.

A rigorous analysis of a preferential attachment model is described in Chapter 4 of [37] and uses the properties of a martingale stochastic process. This method first derives an asymptotic expression of $E[N_{k,t}]$, as $t \to \infty$, and then shows the concentration of $N_{k,t}$ around its mean value $E[N_{k,t}]$, either by using the corresponding properties of martingales or by using variance methods.

5.3.4   Extensions of the Barabási–Albert Model

Variations of Preferential Attachment

The BA algorithm considers only linear preferential attachment, i.e., the probability of connecting to a node depends linearly on its degree. However, other forms of preferential attachment exist with some of them being more representative of the actual attachment taking place in real networks, than the linear one presented before. To be more precise, experimental results in real networks, such as the Internet, the citation network, or the actor collaboration network, show that in some cases the preferential attachment regime is a linear function of the node degree, while in other cases it is sublinear or superlinear [100], depending on the correlations emerging among nodes. In addition, $\Pi(k)$ assigns zero probability to a node with no connections in the network. Such a node, with zero degree, cannot exist in the case of the BA model; however, this situation is possible when nodes or edges are also deleted rather than only being added. Therefore, a form of preferential attachment with "zero-degree attractiveness" or "initial attractiveness" is required [54]. In addition, a node i may have an intrinsic ability to attract links from new nodes expressed by its fitness constant η_i. As proposed in [30], the degree of the node i can be multiplied by the fitness constant of the node η_i. Therefore, a recently added node with small degree can be highly attractive and increase its degree in a fast rate, if its fitness constant is high enough. Following the previous discussion, we categorize below possible forms of preferential attachment.

- Linear preferential attachment:

$$\Pi(x_i) = \frac{x_i}{\sum_{\forall j} x_j} \tag{5.9}$$

- Power-based preferential attachment:

$$\Pi(x_i) = \frac{x_i^a}{\sum_{\forall j} x_j^a} \tag{5.10}$$

  - if $a < 1$ the preferential attachment regime is sublinear, if $a > 1$, it is superlinear and the linear preferential attachment is a special case for $a = 1$.

- Preferential attachment with initial attractiveness:

$$\Pi(x_i) = \frac{A + x_i}{\sum_{\forall j}(A + x_j)} \tag{5.11}$$

  where A is a constant.

- Preferential attachment with fitness:

$$\Pi(x_i) = \frac{\eta_i x_i}{\sum_{\forall j}(\eta_j x_j)} \tag{5.12}$$

where η_i is the fitness constant of node i. The case that $\eta_i = 1$ corresponds to the classic linear preferential attachment regime.

- General form of preferential attachment:

$$\Pi(x_i) = \frac{A + \eta_i x_i}{\sum_{\forall j}(A + \eta_j x_j)} \tag{5.13}$$

By determining the different parameters of Eq. (5.13) as described above, more specific cases of preferential attachment can be obtained and the form that adapts better to each particular occasion can be selected.

But what are the effects in the resulting degree distribution, if the generalized preferential attachment is applied to an evolutionary model similar to the BA's model? It is analytically calculated [100] that the nonlinear preferential attachment ($a > 1$ or $a < 1$) negatively affects the scale-free structure of the network. However, when the preferential attachment is asymptotically linear, i.e., when it tends to a linear function of x_i as x_i tends to infinity, then the network preserves its scale-free nature (with a power-law exponent different than 3). It is also analytically proved [54] that the constant A due to the initial attractiveness does not change the scale-free property of the network structure, but rather only the exponent of the power-law degree distribution. Finally, the fitness constant leads to a power-law degree distribution with a logarithmic correction [30].

Holistic Modification Framework

As mentioned above, the BA model considers only a growing network by ignoring processes such as edge rewiring, edge removal, or node removal. However, in real-world networks, such as online social networks or the World Wide Web, all processes (edge addition/deletion/rewiring, node addition/deletion) may take place. The aim of this section is to present a holistic framework of modeling network evolution and pinpoint special cases from the bibliography with available results regarding the node degree distribution. To begin with, we describe a general framework of network evolution, which can be adapted by appropriately specifying its parameters so as to mimic the evolution of a particular real network.

We categorize the processes constituting the network evolution in two parts denoted as Edge Churn, which consists of edge addition/deletion/rewiring, and Node Churn, which consists of node addition and node deletion. We consider relational graphs, such as peer-to-peer networks, social networks, etc. Therefore all pairs of nodes can be potential neighbors. We denote with $\mathcal{N}(i)$ the direct neighbors of node i and with N_t the number of nodes at time t. Also, for ease of notation we write k_i instead of $k_i(t)$, when this causes no confusion.

- Edge Churn

  – Edge Addition: With probability p_1 we add m_1 new edges between existing nodes. For each new edge $i - j$, the first endpoint, i, is selected with probability $P_1^a(k_i)$ and the second endpoint, j, with probability $P_1^b(k_j)$. This process is repeated m_1 times.

  – Edge Rewiring: With probability p_2 we rewire m_2 existing edges. The node i from which the link is deleted is chosen with probability $P_2^a(k_i)$, while the new endpoint, j, is selected with probability $P_2^b(k_j)$. The endpoint of the rewired link that remains the same does not change its connectivity due to this process. This process is repeated m_2 times.

  – Edge Deletion: With probability p_3 we delete m_3 existing edges. A node i is selected with probability $P_3(k_i)$ and randomly deletes one of its links. This process is repeated m_3 times.

- Node Churn

  – Node Addition: With probability q_1 we add n_1 new nodes, each one creating M new links with older nodes, each one selected with probability $P_a(k_i)$.

  – Node Deletion: With probability q_2 we delete n_2 nodes, along with all their edges, where each node is selected with probability $P_b(k_i)$.

- No action takes place with probability $1 - p_1 - p_2 - p_3 - q_1 - q_2$.

Based on continuum theory, we derive the following expression for the rate of change of the average node degree k_i, $\forall\ i$.

$$\frac{dk_i}{dt} = p_1 m_1 \left[P_1^a(k_i) + P_1^b(k_i) \right] + p_2 m_2 \left[P_2^b(k_i) - P_2^a(k_i) \right] -$$

$$p_3 m_3 \left[P_3(k_i) + \sum_{j \in \mathcal{N}(i)} \frac{P_3(k_j)}{k_j} \right] + q_1 n_1 M P_a(k_i) - q_2 n_2 \sum_{j \in \mathcal{N}(i)} P_b(k_j),\ (5.14)$$

where the right hand side of Eq. (5.14) consists of the sum of the rates of each process separately. Specifically, the first summand corresponds to edge addition, the second to edge rewiring, the third to edge deletion, and the last two summands to node addition and deletion, correspondingly. If only node addition is considered, then the model is identical to the BA model with $M = m$, $q_1 = p$, $P_a(k_i) = \Pi(k_i)$ and $n_1 = 1$ and $P_1 = P_2 = P_3 = q_2 = 0$. In order to understand how the above equation is derived, we explain more explicitly the third summand that corresponds to edge deletion, which can be written in "events" formalism:

{edge deletion happens with probability p_3}∩{node i has m_3 chances of losing a link}∩

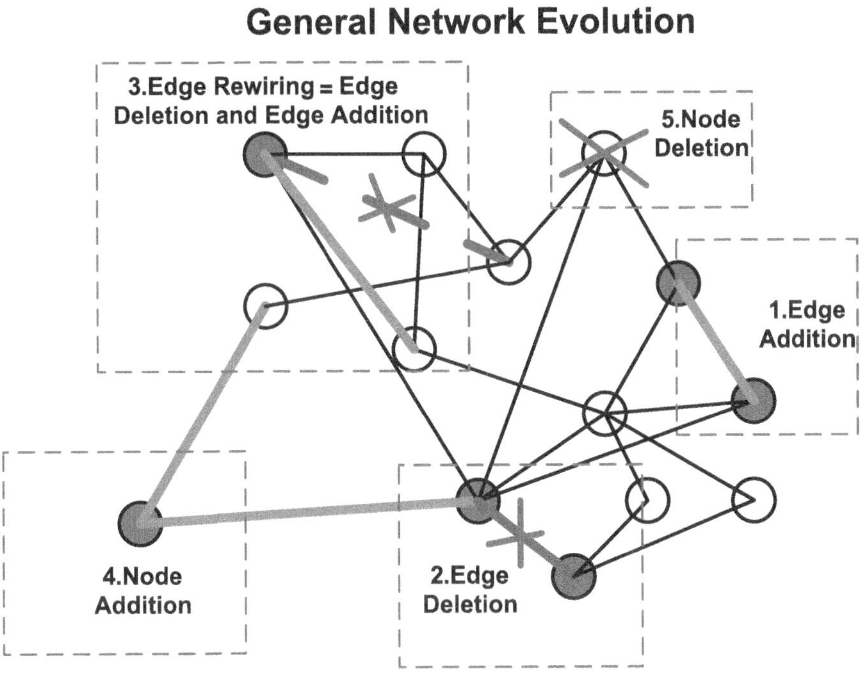

General Network Evolution

Figure 5.7: Visualization of the network evolution in the case of a relational graph. Edge Churn consists of the three processes with IDs 1, 2, 3 while, Node Churn consists of the processes with IDs 4, 5.

{{node i is chosen with probability $P_3(k_i)$ and deletes one of its links}$\cup$ {{one of the neighbors j of node i is chosen with probability $P_3(k_j)$} $\cap$ {j chooses i with probability $\frac{1}{k_j}$, so as to delete the link ji}}}.

The other summands are derived under the same logic. Figure 5.7 illustrates the evolutionary processes taking place in a dynamic relational graph with respect to time evolution.

At this point, we examine a special case of the above framework, which is studied by Barabási–Albert in [5]. This example is restricted to edge addition, edge rewiring, and node addition. Therefore, it consists of the following processes:

- Edge Churn

  – Edge Addition: With probability $p_1 = p$ we add $m_1 = m$ new edges between existing nodes. The first endpoint, i, is selected with probability $P_1^a(k_i) = \frac{1}{N_t}$ and the second endpoint, j, with probability $P_1^b(k_j) = \frac{1+k_j}{\sum_{\forall l}(1+k_l)}$.

- Edge Rewiring: With probability $p_2 = q$ we rewire $m_2 = m$ existing edges. The node from which the link is deleted is chosen with probability $P_2^a(k_j) = \frac{1}{N_t}$ (randomly), while the node where the link will be reconnected is selected with probability $P_2^b(k_j) = P_1^b(k_j) = \frac{1+k_j}{\sum_{\forall l}(1+k_l)}$.

- Node Churn

  - Node Addition: With probability $q_1 = 1 - p - q$ we add $n_1 = 1$ new nodes, each one creating $M = m$ new links with older nodes, each one selected with probability $P_a(k_i) = \frac{1+k_i}{\sum_{\forall l}(1+k_l)}$.

Therefore, the rate equation for the average degree, i.e. Eq. (5.14), takes the form:

$$\frac{dk_i}{dt} = pm\left[\frac{1}{N_t} + \frac{1+k_i}{\sum_l(1+k_l)}\right] +$$
$$qm\left[\frac{1+k_i}{\sum_{\forall l}(1+k_l)} - \frac{1}{N_t}\right] + (1-p-q)m\frac{1+k_i}{\sum_{\forall l}(1+k_l)} \qquad (5.15)$$

which can be written in a more compact form:

$$\frac{dk_i}{dt} = (p-q)m\frac{1}{N_t} + m\frac{1+k_i}{\sum_{\forall l}(1+k_l)} \qquad (5.16)$$

leading to a generalized power-law degree distributions, as shown in [5].

Weighted and Directed Network Graphs

The evolutionary models presented in the previous sections use a network representation as an undirected and binary graph. Realistic networks, however, may consist of directed connections, namely those that are formed unilaterally only by one node towards the other. As an example, in Twitter, a user may unilaterally follow another user, where the latter may not want to reciprocate this connection. On the other hand, weighted graphs are increasingly used nowadays to represent realistic networks, as they do not only encode information regarding connectivity, but also information regarding the properties of each one connection. As an example a weight on a link may correspond to the strength or the cost of the connection or the trust value that the one node poses for the other node. In this section, we are going to extend the BA evolutionary model in the case of weighted and directed graphs [20, 21, 29, 160, 168]. We describe the corresponding extension for weighted networks presented in [18], which is further extended in [19] for weighted and directed networks.

In order to develop an evolutionary model for weighted networks, one should take into consideration, apart from the mechanisms of growth and preferential attachment, the evolution of the weights on the graph if present as well. To be more precise, let us consider the weighted airport network graph (similar in structure to the worldwide airport graph of Section 5.3.2, but extended to bear weights on its links), with weights corresponding to the number of passengers transferred between two directly connected airports. Then, the addition of a new connection to an airport will increase the number of passengers traveling out of this airport, therefore it will increase the weights of the other links of the particular airport and this weight change will be transmitted to other connected airports as a chain. The same weight evolution takes place on the weighted graph of the Internet routers, where now the weights represent traffic. However, if the weights represent trust values, this weight evolution may not have a tangible meaning. The model developed by Barrat, Barthélemy, and Vespignani in [18] considers for simplicity only local weight evolution, i.e., the addition of a link to a node affects only the weights of the links connected to this node. In the analysis that follows, we use $\mathcal{N}(i)$ to denote the one-hop neighborhood of node i. This evolutionary model considers a connected initial core network of N_0 nodes and homogeneous link weights equal to w_0 and consists of the following processes:

- Growth: At each time step t, a new vertex with ID t is added and connects to m already existing vertices, each of which is selected according to the rule of strength-driven preferential attachment, i.e., with probability defined as: $\Pi(s_i) = \frac{s_i(t)}{\sum_{\forall j} s_j(t)}$, where $s_i(t) = \sum_j w_{ij}(t) = \sum_j w_{ji}(t)$ (due to the assumption of an undirected network graph, also see Chapter 2). The probability $\Pi(s_i)$ assigns higher priority of selection to nodes central in terms of strength of interactions. The weight of each new edge is set to w_0.

- Weight Evolution: The addition of a new edge to node i, at time step t, with weight w_0 will cause a perturbation of the weights on the rest of the links with endpoint at the node i, which, for link (i, j) equals $\Delta w_{ij}(t)$ and is described by the rule: $w_{ij}(t) \leftarrow w_{ij}(t) + \Delta w_{ij}(t)$. The proposed model assumes that each new link added at node i induces an increase equal to δ_i on s_i which is distributed to the weights $w_{ij}(t)$ for all nodes $j \in \mathcal{N}(i)$, proportionally with their value, i.e.: $\Delta w_{ij}(t) = \delta_i \frac{w_{ij}(t)}{s_i(t)}$. Therefore, the strength of node i is adapted as $s_i(t) \leftarrow s_i(t) + w_0 + \delta_i$.

According to the above model, we can obtain analytical expressions for the variables s_i, k_i, w_{ij} based on continuum theory, following the same approach as the BA model, i.e., by considering s_i, k_i, w_{ij} as continuous variables. Therefore, we obtain the following differential equations:

$$\frac{dk_i(t)}{dt} = m\Pi(s_i) = m\frac{s_i(t)}{\sum_{\forall l} s_l(t)}, \tag{5.17}$$

which is similar to the BA model,

$$\frac{ds_i(t)}{dt} = m\frac{s_i(t)}{\sum_{\forall l} s_l(t)}(1+\delta_i) + \sum_{j\in\mathcal{N}(i)} m\frac{s_j(t)}{\sum_{\forall l} s_l(t)}\delta_j\frac{w_{ij}(t)}{s_j(t)}, \qquad (5.18)$$

where at the right hand side, the first term corresponds to the increase of s_i because of the selection of node i with strength-driven preferential attachment, while the second summand is due to the increase in the strength of node i because of the selection of a neighbor of i,

$$\frac{dw_{ij}(t)}{dt} = m\frac{s_i(t)}{\sum_{\forall l} s_l(t)}\delta_i\frac{w_{ij}(t)}{s_i(t)} + m\frac{s_j(t)}{\sum_{\forall l} s_l(t)}\delta_j\frac{w_{ij}(t)}{s_j(t)}, \qquad (5.19)$$

where at the right hand side, the first term corresponds to the increase in w_{ij} due to the selection of node i, while the second summand is due to the selection of the endpoint j of the link ij.

We solve the above equations in the simple case where $\delta_i = \delta, \forall\ i$. This case is denoted by the authors in [18] as homogeneous coupling, and considers that the addition of a link with weight w_0 to node i causes a node-independent weight increase to the strength of node i, equal to δ. However, this is a theoretical model, a more realistic version of which is analytically described in [18], where parameter δ_i is different for each node i. This model is denoted as heterogeneous coupling. From now on, we consider all the weights normalized by the initial weight w_0. A new edge to an existing node adds weight to the entire network, equal to $2(1+\delta)$. As a result, $\sum_{\forall l} s_l(t) \approx 2(1+\delta)mt$ (if we ignore the initial sum of weights in the network graph) and similarly $\sum_{\forall l} k_l(t) \approx 2mt$. With respect to homogeneous coupling, Eq. (5.18) can be written as:

$$\begin{aligned}
\frac{ds_i(t)}{dt} &= m\frac{s_i(t)}{\sum_{\forall l} s_l(t)}(1+\delta) + \sum_{j\in\mathcal{N}(i)} m\frac{s_j(t)}{\sum_l s_l(t)}\delta\frac{w_{ij}(t)}{s_j(t)} \\
&= m\frac{s_i(t)}{2(1+\delta)mt}(1+\delta) + m\frac{s_i(t)}{2(1+\delta)mt}\delta = \frac{s_i(t)}{2(1+\delta)mt}m(1+2\delta)
\end{aligned}$$

$$\qquad (5.20)$$

$$\Rightarrow\ s_i(t) = m\left(\frac{t}{i}\right)^{\frac{2\delta+1}{2(\delta+1)}}, \quad if\ s_i(t=i)=m. \qquad (5.21)$$

where, $t=i$ is the time on which node i entered the network. By using Eq. (5.20), Eq. (5.17) takes the following form:

$$\frac{dk_i(t)}{dt} = m\Pi(s_i) = m\frac{s_i(t)}{\sum_l s_l(t)} \qquad (5.22)$$

$$\Rightarrow k_i(t) = \frac{s_i(t)+2m\delta}{2\delta+1}, \qquad (5.23)$$

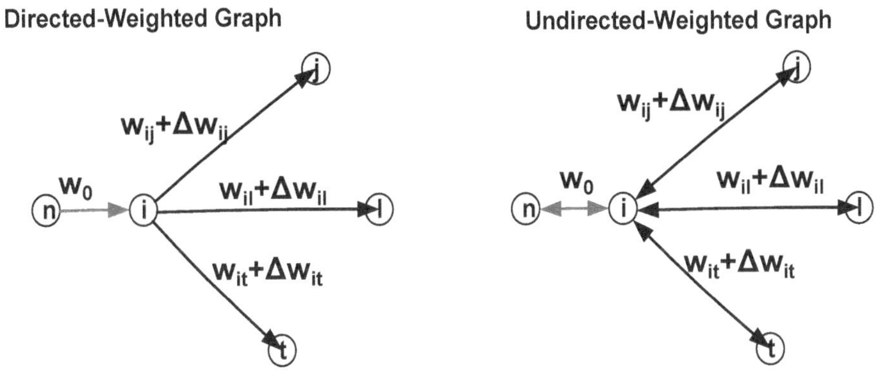

Directed-Weighted Graph   **Undirected-Weighted Graph**

Figure 5.8: Extension of the Barabási–Albert Model for the evolution of weighted and directed graphs.

which correlates the degree of node i with its strength. Finally, Eq. (5.19) is transformed to:

$$\frac{dw_{ij}(t)}{dt} = m\frac{2}{\sum_l s_l(t)}\delta w_{ij} = m\frac{2w_{ij}(t)}{2(1+\delta)mt} = \frac{\delta w_{ij}(t)}{(1+\delta)t} \tag{5.24}$$

$$\Rightarrow w_{ij}(t) = \left(\frac{t}{t_{ij}}\right)^{\frac{\delta}{\delta+1}}, \tag{5.25}$$

where t_{ij} is the time of appearance of the link ij and $t_{ij} = \max(i,j)$.

Therefore, for all the above metrics (node degree, strength, and link weights) analyzed, the yielded topologies develop power-law time relations.

Figure 5.8, shows the addition of a new link between a newcomer node n and node i, and the weight adaptation due to the increase of the strength of node i. Both cases of weighted undirected graphs and weighted directed graphs are depicted.

A similar procedure is followed for the construction and the analysis of evolutionary models for both weighted and directed network graphs. However, contrary to the case of an undirected network graph, in a directed network one should model separately the evolution of the in-degree, $k_i^{in}(t)$, the out-degree, $k_i^{out}(t)$, the in-strength, $s_i^{in}(t)$, and the out-strength, $s_i^{out}(t)$. In [19], the authors study an evolutionary model corresponding to the evolution of the weighted and directed World Wide Web Graph. The vertices correspond to the Web pages, the directed links correspond to hyperlinks from one Web page to another, and the link weight corresponds to the number of users (traffic) using the corresponding hyperlink. They prove that $k_i^{in}(t), s_i^{in}(t), s_i^{out}(t)$ follows a power-law degree distribution, while for their model, $k_i^{out}(t)$ is constant by construction.

To briefly sketch the model described in [19], it is very similar to the evolutionary model of [18] described just above; however, when node i is added at time step $t = i$, it has an out-degree $k_i^{out}(i) = m$. Thus it attaches through directed connections (hyperlinks) to m already existing nodes (Web pages) increasing their in-degree. Each one of the m existing nodes is selected with probability $\Pi(s_i^{in}) = \frac{s_i^{in}(t)}{\sum_j s_j^{in}(t)}$, at time t, due to the fact that $s_i^{in}(t)$ expresses node i' s popularity. Each new directed edge is assigned a weight of w_0 and causes a weight adaptation to the node that points to, in similar manner as in [18]. The out-degree $k_i^{out}(t) = m$ remains for all t constant as a node obtains only new in-connections through the addition of new nodes. The analytical equations of $k_i^{in}(t), s_i^{in}(t), s_i^{out}(t)$ are obtained through continuum theory in the same fashion and following the same logic as described above for the model of [18].

5.4   Hyperbolic Structure of Complex Networks

In the previous sections, we have described analytically some of the most characteristic features of the structure of complex networks, i.e., the small-world paradigm and the scale-free property along with the evolutionary models that mimic the dynamics of real complex networks' evolution and lead to network models with these particular properties. To sum up the basic evolutionary models, the Watts and Strogatz model constructs a small-world network while the Kleinberg model leads to a small-world navigable network, i.e., not only the final network graph has a small-world structure but the nodes can actually find the short paths without using global information of the network topology. Finally, the Barabási–Albert model uses the preferential attachment rule and the growth mechanism in order to produce networks with power-law degree distributions, i.e. scale-free networks.

In this section, we study another category of evolutionary models that leads to network topologies that appear to emerge both scale-free and small-world properties. This new category is based on a very useful observation: the hyperbolic hidden metric space that seems to exist behind the complex networks' structure [101], [32]. Initially, we provide some basic knowledge about the hyperbolic geometry and the two dimensional hyperbolic plane, which will be exploited for modeling purposes in the sequel. Also, we give an intuitive explanation about the connection between the hyperbolic metric space and scale-free complex networks [124].

5.4.1   Background on Hyperbolic Geometry

The whole infinite hyperbolic plane can be represented inside the finite unit disk $\mathbb{D} = \{z \in \mathbb{C} | |z| < 1\}$ of the Euclidean space. Hyperbolic geometry has negative curvature (contrary, i.e., to the spherical geometry), as mentioned above and analyzed in Section 4.7. This model of visualization of the hyperbolic space, where nodes are represented with complex numbers in the set $\mathbb{D}$, is denoted as the Poincaré Disk model. However, there are other models for visualization of a hyperbolic geometry space, a summary of which can be found in [8]. The hyperbolic distance function $d_H(z_i, z_j)$, for two points z_i, z_j, in the Poincaré Disk model is given by:

$$\cosh d_H(z_i, z_j) = \frac{2|z_i - z_j|^2}{(1 - |z_i|^2)(1 - |z_j|^2)} + 1 \qquad (5.26)$$

The Euclidean circle $\partial\mathbb{D} = \{z \in \mathbb{C} | |z| = 1\}$ is the boundary at infinity for the Poincaré Disk model. In addition, in this model, the shortest hyperbolic path between two nodes is either a part of a diameter of $\mathbb{D}$, or a part of a Euclidean circle in $\mathbb{D}$ perpendicular to $\partial\mathbb{D}$. Figure 5.9 shows a network embedded inside the Poincaré Disk.

In the following, we explain the similarity between an infinite tree graph and the hyperbolic space, aiming to provide an intuition about the hidden hyperbolic structure of complex networks. Considering the two-dimensional space with constant curvature -1, the length of the circle and the area of a disk of radius R are $2\pi \sinh R \sim e^R$ and $2\pi(\cosh R - 1) \sim e^R$, correspondingly, which are exponentially growing with the radius R. This exponential scaling coincides with the scaling of the number of nodes with respect to their distance form the root of the tree in an "e-ary" tree, as stated in [124]. To make this clearer, let us examine a b-ary tree, which is a tree with branch factor equal to b. In the case of the b-ary tree, the number of nodes located at distance exactly R from the root of the tree is $(b + 1)b^{(R-1)} \sim b^R$ and the number of nodes being at distance at most R from the tree is $\frac{(b+1)b^R - 2}{(b-1)} \sim b^R$. As a result of the described similarity of scaling between the tree and the hyperbolic space, the hyperbolic space can be seen as a continuous version of a tree and this is denoted as the exponential expansion property of the hyperbolic space. Scale-free complex networks are characterized by heterogeneity regarding the node degree, leading to the categorization of the nodes in groups and subgroups based on their node degree, which can be represented by a tree, and this observation, as a result, implies the existence of a hidden hyperbolic metric space.

Figures 5.9 and 5.10 serve the purpose of illustrating the form of some basic hyperbolic geometric objects. Figure 5.9, depicts a tree-network embedded in the Poincaré Disk Model, where two neighboring nodes are connected via their shortest paths. Figure 5.10(a) shows two circles inside the Poincaré Disk Model, where it can be observed that the center is not located in the middle of the circle as it happens in the Euclidean metric space. Figure 5.10(b) shows a hyperbolic polygon with the non-straight shortest paths connecting two consecutive points. Finally, Figure 5.10(c), depicts a tessellation of the Poincaré Disk into hyperbolic triangles illustrating how distances scale in the hyperbolic space as we move closer to the infinite boundary $\partial\mathbb{D}$.

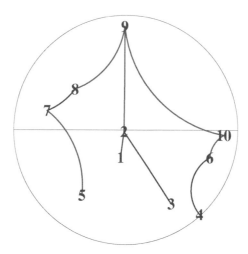

Figure 5.9: Visualization of a tree network embedded in hyperbolic coordinates. The edge between two neighbors represents their hyperbolic shortest path. As stated in the text, it is either part of a diameter in $\mathbb{D}$, i.e., edge $(2,9)$ or part of a Euclidean circle in $\mathbb{D}$ perpendicular to $\partial\mathbb{D}$, i.e., edge $(7,5)$ or $(6,4)$. (Figure created with [157].)

5.4.2 Evolutionary Models Developed on Hyperbolic Geometry

In this subsection, we are going to present the network growing model introduced in [124] for the hyperbolic metric space. The created network has a power-law degree distribution, it is small-world navigable through greedy routing and is highly clustered. We are mostly focused on the general philosophy followed by the model and its outcomes, rather than its exact algorithmic realization. We assume polar coordinates for the nodes (r, θ). A uniform distribution of N nodes on the hyperbolic plane over a disk of radius R is performed by assigning to each node an angular coordinate randomly and uniformly chosen in the interval $[0, 2\pi]$ and a radius coordinate $r \in [0, R]$ chosen with exponential density $f(r) = \frac{sinh r}{(cosh R - 1)} \approx e^{(r-R)} \sim e^r$, due to the exponential growing length of a circle with the radius r. For achieving a non-uniform distribution of nodes, we use a parameter a in $f(r)$ as $f(r) = \frac{a sinh ar}{(cosh aR - 1)} \approx a e^{a(r-R)} \sim e^{ar}$. Initially the network is considered empty. At each time slot $t = i$, a node with ID i enters the network and after defining its polar coordinates, connects to some appropriately chosen nodes as it is described below. In addition, at each time slot, the disk area over which the nodes are placed increases. At each time slot $t = i$, the new node i performs the following:

- It selects its angular coordinate θ randomly and uniformly inside the interval $[0, 2\pi]$.

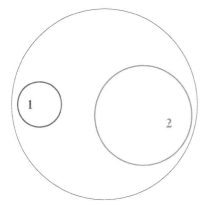 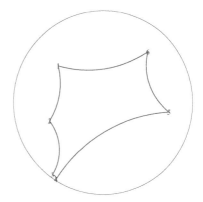

(a) Two hyperbolic circles in the Poincaré Disk Model. The first one has as center the point with label 1 and radius 1.2, while the second has as center the point with label 2 and radius 2. (Figure created with [157].)

(b) A hyperbolic polygon in the Poincaré Disk Model, where the lines connecting two points are identified with their hyperbolic shortest paths. (Figure created with [157].)

(c) Tessellation of Poincaré Disk Model into hyperbolic triangles, showing the exponential scaling in the hyperbolic geometry.

Figure 5.10: Demonstration of geometric schemes in the Poincaré Disk Model.

- It computes the new increased hyperbolic disk radius as $R(i) = \frac{1}{a} \ln \frac{i}{c}$.

- It selects its radial coordinate $r \in [0, R(i)]$ according to the distribution $f(r|R(i)) = \frac{a \sinh ar}{(\cosh aR(i)-1)} \approx ae^{a(r-R(i))}$.

- It connects to every other node j with hyperbolic distance from i less than $R(i)$.

The above model, which links node pairs without the explicit use of preferential attachment, leads to power-law degree distribution, i.e., the probability of node degree k at time t is $P(k,t) \sim k^{-\gamma}$, with exponent $\gamma = 2a+1$, $a \geq \frac{1}{2}$.

In addition, the networks created by the above model appear to be highly clustered since the connections with close nodes in hyperbolic distance lead to the formation of a large number of triangles in the network structure. Finally, the networks with hidden hyperbolic structure perform very well under greedy routing based on hyperbolic distances, as it is shown through experimental examination. The term "good performance" denoted the fact that the paths followed by the packets through greedy routing are very close to the global shortest paths between the corresponding node pairs. This is a very important property showing the small-world navigability of this particular category of networks. The nodes discover the global shortest paths by using only local information consisting of their coordinates, the coordinates of their one-hop neighbors, and the coordinates of the destination the packet is intended for.

In [123], a special case of the above model is examined, where the polar coordinates of a node i are assigned as a function of the time that i entered the network and more precisely $r_i = \ln i$ and the angular coordinate expresses the preferences of i. Then the difference in angular coordinates between two nodes corresponds to their similarity and the polar coordinate of a node reflects its age and thus its popularity, if considering the older nodes as more popular. In this case, the hyperbolic distance between two points at polar coordinates (r_1, θ_1), (r_2, θ_2), which is approximated by the type $d_{12} = r_1 + r_2 + \ln(\frac{\theta_{12}}{2})$, where $\theta_{12} = \theta_1 - \theta_2$, expresses the combination of the two metrics of popularity and similarity between the two points. Therefore, the connections based on hyperbolic distance are determined through the popularity and the similarity metrics of the candidate node pairs. For this special case, the probability that a node of degree k is selected for a connection by a newcommer is computed and found to be equal to the preferential attachment probability $P(k)$ (Section 5.3.4).

5.5 Expansion Properties

5.5.1 Definition and Analytical Properties

Expander graphs constitute a field of spectral graph theory with interesting and important applications in many directions including hardness problems, error correcting codes, algorithmic design, and analysis of complex networks. In this book, we are mostly interested in the expansion properties of complex communication networks and their practical significance. Specifically, the expansion of a graph is a parameter that measures the degree to which a graph is simultaneously sparse and highly connected, which is by definition a counter-intuitive notion. In the following, we provide the definition and the basic analytic properties of expander graphs in mathematical formalism, while in the next section, we illustrate their use by describing possible applications.

Let us denote as $\bar{S}$ the complement of a set S of vertices in a graph $G = (V, E)$ with $|V| = n$ nodes, i.e., $\bar{S}$ includes the vertices of V that do not belong to S (Section 2.1). We denote with $\vartheta S = E(S, \bar{S})$ the "edge boundary" of the set S, which is defined as the set of edges with exactly one endpoint inside set S and the other endpoint inside $\bar{S}$. The expansion of a set S with $|S| \leq \frac{n}{2}$ is defined as $\frac{|\vartheta S|}{|S|}$. The expansion ratio of a graph G is defined as the least expansion of any set $S \subset V$, where $|S| \leq \frac{n}{2}$, as in Definition 20 [89].

Definition 20 *The expansion ratio of a graph G, based on the edge boundary, is defined as* $h(G) = min_{\{S||S|\leq\frac{n}{2}\}} \frac{|\vartheta S|}{|S|}$.

It is important to notice that the expansion ratio of a graph is the minimum expansion of every set $S \subseteq V$ with $|S| \leq \frac{n}{2}$. This means that when sampling a set of nodes S from G with less that $\frac{n}{2}$ nodes, it is ensured that there are at least $h(G)|S|$ connections beginning inside the sampled set S and ending in $\bar{S}$. As an example, the Petersen graph (Figure 5.11) has $h(G) = 1$ (according to Definition 20) and similarly for the complete graph with n nodes, K_n, each subset S with $|S| = l \leq n$ will have $l \cdot (n - l)$ edges in its edge boundary and for $l = \frac{n}{2}$, $h(G)$ takes its lower value equal to $n - \frac{n}{2} = \frac{n}{2}$.

Another definition of the expansion ratio is obtained if instead of considering the edge boundary of the set $S \subseteq V$, we consider the Node Boundary of the set S. As Node Boundary of S, we define the set of nodes $N(S)$, where $N(S) = \{w \in V - S : \exists u \in S \; s.t. \; (u, w) \in E\}$.

Definition 21 *The expansion ratio of a graph G based on the node boundary, is defined as* $h'(G) = min_{\{S||S|\leq\frac{n}{2}\}} \frac{|N(S)|}{|S|}$.

In general, the computation of the expansion ratio of a graph is NP-hard [89]; however, there exist spectral methods that simplify the computations due to the existence of key relations between the spectrum of a graph

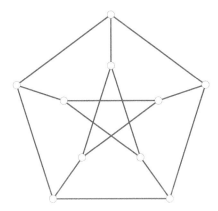

Figure 5.11: Visualization of Petersen graph.

and its expansion properties. More precisely, the expansion ratio is highly correlated with the spectrum of the graph, i.e., the set of the eigenvalues of the adjacency matrix A of the graph, through some basic theorems of the expander graphs' analysis. In the sequel, we briefly present some definitions, theorems, and applications for d-regular graphs with symmetric adjacency matrix $n \times n$ (undirected graphs), in order to show briefly how the expansion properties are connected to the spectrum of the graph (see also Section 2.1.8). Since the matrix A is real and symmetric, it has n real eigenvalues. We consider that the eigenvalues of the adjacency matrix A of the graph G are ordered as $\lambda_1 \geq \lambda_2 \geq \lambda_3 ... \geq \lambda_n$.

Definition 22 *A sequence of d-regular graphs $\{G_i\}_{i \in \mathbb{N}}$ of size increasing with i is a family of expander graphs, if there exists an ϵ so that $h(G_i) > \epsilon$ for each i.*

Intuitively, every small subset of an expander graph has a large neighborhood. In the case of d-regular graphs, the largest eigenvalue, λ_1, is equal to d, i.e., $\lambda_1 = d$.

Definition 23 *The spectral gap of a d-regular graph is defined as $SG = d - \lambda_2$, expressing the difference between the first and the second eigenvalue.*

The following theorem connects the spectral gap of a d-regular graph with its expansion ratio.

Theorem 58 *Let G be a finite d-regular graph on n nodes with spectrum $d \geq \lambda_2 \geq \lambda_3 ... \geq \lambda_n$. Then, $\frac{d - \lambda_2}{2} \leq h(G) \leq \sqrt{2 \cdot d \cdot (d - \lambda_2)}$.*

The main observation derived from Theorem 58 is that good expander graphs have high spectral gap. This theorem is also extended to general graphs by replacing d with λ_1 and the difference $\lambda_1 - \lambda_2$ of the first and second eigenvalues of the adjacency matrix of a network graph defines the spectral gap of the corresponding graph. In this case if $\lambda_1 - \lambda_2$ is high, then the network is considered a good expander. However, according to the following Alon–Boppana theorem, λ_2 is lower-bounded, meaning that the spectral gap of a graph cannot increase infinitely. Let us define as $\lambda = \max\{|\lambda_2|, |\lambda_n|\}$.

Theorem 59 *(Alon–Boppana): Let G be a finite d-regular graph with n vertices. Then, $\lambda \geq 2\sqrt{d - 1} - o_n(1)$, where $o_n(1)$ is a quantity that for fixed d tends to 0 as $n \to \infty$.*

The use of the spectral gap $\lambda_1 - \lambda_2$ as a metric for determining if a particular graph is an expander or not has the drawback that we do not know how large it should be so as to accurately decide. In order to surpass this difficulty, we present another way of examining and characterizing the expansion properties [61] of any general graph that links a centrality measure denoted as subgraph centrality [62] with the first eigenvalue of the adjacency matrix of a network graph. The definition of the subgraph centrality of node i involves the number of closed walks that start and end at node i. More precisely, the subgraph centrality of i [62] is defined as:

$$Sc(i) = \sum_{k=0}^{\infty} \frac{\mu_k(i)}{k!}, \qquad (5.27)$$

where $\mu_k(i) = A^k(i,i)$, i.e., the (i,i) entry of the k-th power of the adjacency matrix A, which coincides with the number of closed walks starting and ending at node i, with length equal to k. If we consider only closed walks with odd length (since a cycle of even length may correspond to back and forth movements on acyclic subgraphs) for non-bipartite graphs (for bipartite graphs, the number of cycles with odd length is zero (Theorem 1)) and by using the spectral graph theory, the subgraph centrality of node i can be written as follows:

$$Sc_{odd}(i) = \sum_{j=1}^{n} [\gamma_j(i)]^2 \sinh(\lambda_j) = [\gamma_1(i)]^2 \sinh(\lambda_1) + \sum_{j=2}^{n} [\gamma_j(i)]^2 \sinh(\lambda_j),$$

$$(5.28)$$

where $\lambda_j, \gamma_j(i)$ are the jth eigenvalue and the ith component of the jth eigenvector, correspondingly. In the case of expander graphs, as mentioned above, the spectral graph difference is large, thus $\lambda_1 \gg \lambda_2 \geq \lambda_3... \geq \lambda_n$ and the subgraph centrality of node i can be approximated as follows:

$$Sc_{odd}(i) \approx [\gamma_1(i)]^2 \sinh(\lambda_1) \Rightarrow \gamma_1(i) \propto \sinh(\lambda_1)^{-\frac{1}{2}} Sc_{odd}(i)^{\frac{1}{2}}, \qquad (5.29)$$

which relates directly the principal eigenvector (eigenvalue) with the subgraph centrality and leads to a linear relation in log-log scale:

$$\frac{1}{2} \log Sc_{odd}(i) - \frac{1}{2} \log \sinh(\lambda_1) = \log \gamma_1(i). \qquad (5.30)$$

As a result, a graph is an expander, if a log–log plot of $\gamma_1(i)$ versus $Sc_{odd}(i)$, $\forall\, i = \{1..n\}$ is linear with slope approximately equal to $\frac{1}{2}$.

At this point, we present some experimental results based on the above analysis from [61] regarding the expansion properties of complex networks. According to these results, networks such as the Internet (at the autonomous systems level (AS)), the food chain, the bibliographic citation networks, and the US airport transportation network present good expansion properties since they appear to have the desirable scaling relation between the subgraph centrality and the principal eigenvector, while other networks such as the network of the inmates in prison, the injecting drug users, and the protein–protein interaction network do not seem to be good expanders from the experimental results. In [110], it is shown through the same method, i.e., by using the subgraph centrality, that large social networks (like Facebook, Youtube, Epinions, etc.) exhibit very good expansion properties. It is very interesting to point out that, as it is shown in [74] and supported by [61], the scale-free networks that obey power-law degree distributions are expander graphs. As stated in [61], very clustered networks without the existence of shortcuts, such as the network of the drug users, are not expected to be good expanders. A possible procedure for the creation of graphs with good expansion properties would be to create clustered networks with high average node degree and then rewire edges that connect nodes inside the same clusters so as to connect nodes from different clusters (and therefore, the rewired edges are functioning like shortcuts). As a result, in order for a network to be a good expander, it should have higher number of intra-cluster links than inter-cluster links. For this reason, community-based small social networks, such as the network of the drug users or the protein–protein interaction network, do not seem to have expansion properties.

5.5.2   Applications of Expander Graphs

As we mentioned before, the applications of expander graphs are important and extend to many fields. In the sequel, we enumerate and briefly describe some of these applications.

- Random Walk: A simple random walk on an expander graph is similar to choosing vertices uniformly and randomly. This is due to a theorem denoting that the stationary distribution of a random walk on an expander graph is the uniform distribution [89]. Therefore, we can sample a network that is a good expander, uniformly and randomly, more efficiently, i.e., by using less random bits, if performing a random walk on its graph.

- Routing through shortest paths: It can be proved that between two nodes of an expander graph, there is a path, the length of which depends on the expansion ratio of the graph and specifically, as the expansion ratio increases, the path becomes shorter. Through this application, the relation existing between expander graphs and small-world graphs becomes obvious.

- A graph with high expansion ratio has very good communication properties by avoiding bottlenecks. A bottleneck is a set of nodes that if being removed separates the rest of the graph to at least two large components (Chapter 2). The existence of bottlenecks is characteristic of the clustered networks that are not good expanders. Expander Graphs present robustness to node and link failures due to the lack of bottlenecks.

- Error correcting codes: The graph codes defined on expander bipartite graphs can be characterized by a large number of codewords with a large distance between them (robustness to channel errors) and also can be efficiently decoded.

- Complexity: Expander graphs are proven to amplify the success probability of randomized algorithms.

The interested reader may refer to [89] for a more extensive and detailed presentation of expander graphs. Since we are studying communications networks in this book, we are interested in the second and third applications of the expander graphs presented before. In general, by proving that a network graph is an expander, we have proved that it inherently has short paths between the node pairs without being overcrowded with shortcuts, as in the case of small-world networks. Also, routing becomes very efficient by avoiding bottlenecks. Finally, since according to the first application it is possible to select vertices uniformly and randomly by just performing a random walk on the expander graph, in the case of a communication network with expander graph structure, it will be possible to perform very efficiently random node election processes [3], etc. To conclude, proving the existence of the expanding graph structure is hard, but it if it is achieved, it indicates networks that achieve efficiently very good performance. A possible, but difficult direction for future work is to develop mechanisms for emerging the expander graph structure in communication networks.

5.6 Conclusions

In Table 5.3 we provide a summary of the most important structure types identified in complex networks. Since each type of structure characterizes or is characterized by particular trends with respect to a set of evaluation metrics (such as average path length, clustering coefficient, curvature, degree distribution), we provide these trends in Table 5.3 and columns $2-5$. If a trend is not known (or it is not of particular interest for the applications presented in this book) for a specific combination of structure type and metric, the corresponding cell shows an "N/A" sign. In the case of hyperbolic structure, the high clustering refers to the network constructed and studied by the models of [124] and [123], which, however, might not always be the case. Finally, it is important to note that some networks may present features for more than one structure, i.e., the airport network [79] is both scale-free and small-world and similarly the network constructed by [124] combines features from hyperbolic, small-world, and scale-free structures.

Table 5.3: Summary of the most important structure types in complex and social network analysis and their characteristics.

Type of structure	Metrics				Evolutionary construction
	Average path length	Clustering	Curvature	Degree Distribution	
Small-world	Low (like random graphs)	High (like regular lattices)	N/A	N/A	Watts and Strogatz model and Kleinberg's model
Scale-free	Usually low	N/A	Negative	Power-law	Barabási-Albert model and model of [124]
Hyperbolic	Low	High[a]	Negative	Power-law	Model of [124] and [123]
Expander	Low	Low	N/A	N/A	N/A

[a] As observed in the network constructed by [124].

Chapter 6

Evolutionary Approaches

The non-trivial and sometimes rather desired properties of (online) social networks have attracted the interest of the research community (Chapter 5), eventually rendering (online) social networks a new, interesting and evolving research field. Such research interest in the properties and dynamic behavior of social networks is dictated by the fact that some of their salient features can also be observed, exploited, or even incorporated in other more complex network contexts. The most typical example of the latter is the fact that, due to the small-world phenomenon, the communication through a social network can be expedited, as the small-world shortcuts may contract the average distance[1] separating two nodes. In other words, the communication speed is a salient characteristic of social networks and by mimicking the way it is achieved in a social network context, it can be incorporated in other network types as well. Intuitively, by introducing appropriate shortcuts in another network class, one can obtain a network structure and function similar to the small-world phenomenon and this makes it possible to reduce the distance between a communicating pair and leads to an increase of the speed of communication, without modifying the initial character of the network. Many works in the past have established themselves in the bibliography for attempting to apply social network concepts in order to make more efficient the function of various wired or wireless networks. This topic will also be the main focus and objective of the current chapter.

More specifically, in this chapter, we are going to describe and analyze evolutionary approaches mainly applied, but not limited, to wireless multi-hop networks, which are briefly described in Section 6.1. A common viewpoint that can be currently identified in the literature is the fact that the evolutionary approaches regard the physical topology of the wireless network, which is altered in a nominal way so that the network as a graph is enhanced with social features. As a result, the main focus of this chapter is the reconsideration and

---

[1]In the sense defined in each case, e.g., hop-wise.

sometimes even efficient re-design of the wireless multi-hop network topology, without, however, altering its original characteristics as these are defined by the wireless nodes' properties. Both the design procedure and the desirable characteristics are inspired by corresponding methodologies and features of social networks' evolving topology graphs. The improvement of the network performance is either analytically proved beforehand, or observed and quantified after the development through simulations. In addition, the majority of the described mechanisms focus on the small-world property of social networks, and consist of topology modification methods applicable to wireless ad hoc networks that enrich them with small-world features.

This chapter is organized as follows. The topology control mechanism that is the basic tool for the topology modifications applied in wireless multi-hop networks is briefly explained in Section 6.2. Sequentially, in Section 6.3 we discuss the emerging difficulties in the effort of evolving the spatial wireless multi-hop networks into small-world graphs and explain on which level such a modification is attained throughout the presented bibliography. Afterwards, in Section 6.4 we analytically present some of the mechanisms that have been developed for enhancing the wireless ad hoc networks with social features and further make them small-world-like graphs. In the sequel, we focus on the description and analysis of a holistic topology modification approach for enhancing the physical topology of wireless multi-hop networks with small-world features. The mechanism presented in Section 6.5 is of a probabilistic nature, it is reversible in the sense that nodes can revert quickly to their original states, it applies to both weighted and binary graph network models, and eventually it can be adapted properly with respect to the given cost budget, the desirable performance objectives, and the given application requirements.

6.1   A Brief Description of Wireless Multi-hop Communications

In this Section, we briefly present basic elements of wireless multi-hop networks so as to facilitate the flow of the rest of the chapter. We summarize the operation and modeling of the wireless multi-hop networks and we present some metrics that are used to quantify their performance. Such metrics will be exploited in this chapter for assessing the improvemens in performance obtained by the presented mechanisms. Finally, we conclude the section with some references and terminology. We note that this section can be omitted by a reader already familiar with the field of wireless communications, in order to focus on the less well-known material following.

Wireless communications are continuously increasing their penetration in the whole system of networking, basically due to their ability for self-organization and low cost deployment. From a net perspective, a wireless system consists of nodes that communicate with each other through a wireless channel. Some of these networks consist of a backbone wired infrastructure.

As an example in the case of cellular networks, a wireless client communicates directly through the wireless channel with the base station and the latter is connected directely to the wired backbone infrastructure. In this chapter, we are focusing on completely distributed wireless networks such as wireless ad hoc and wireless sensor networks, which are generally denoted by the term wireless multi-hop networks. In this case there is no central authority dictating the nodes' actions. On the contrary, nodes self-organize their topology and functionalities so as to ensure connectivity and good quality of communication. The lack of centralized control and high cost of deployment led to their use in cases where there is no possibility of establishing wired communications (i.e., battlefields, disaster recovery, etc.), or in cases when a temporary infrastructure is needed (i.e., conferences) or we cannot afford the cost of wired networks (i.e., developing countries). In these networks, cooperation among nodes is necessary for achieving the basic network functionalities such as message transmissions and routing.

We assume that the wireless nodes are located either uniformly and randomly (Appendix A) or in a specified topology over a bounded region. Each node i has a limited transmission power and therefore cannot communicate directly with all nodes in the network region, but rather only with those that are located close enough to node i. The communication with the long-distance nodes is achieved via multi-hop transmissions through intermediate nodes acting as routers. Typically, the Random Geometric Graph (RGG) model is used to represent the topology of a wireless multi-hop network. In the RGG model (as also employed in this chapter) each node i is characterized by a transmission radius, i.e., R, and can transmit directly to all nodes j lying on a disc centered at i with radius R. This disc is defined as the communication range of node i and each node j lying in this range is denoted as one-hop (or direct) neighbor of node i. Similarly node i can directly receive a transmission only from a node lying inside its range. The ij (or sometimes denoted as (i,j)) direct communication is denoted as link or edge. The links and nodes compose the corresponding network graph of the wireless multihop network. The term "unicast link" denotes the communication between a single sender and a single receiver, also referred to as end-to-end communication. The network topology and the transmissions define the physical layer of the wireless network function. If i is not the destination of the packets arriving at i, these can be buffered at i until they are further transmitted, when possible, towards the destination. We use the term "when possible" since, as the wireless nodes share the same channel, they may interfere with each other. As a result, two more network functionalities occur here, the data link layer scheduling process, which determines when a node can attempt a physical transmission, and the network layer routing process determining the sequence of nodes that a packet follows from the source to the destination (i.e., the path of the packet). Routing and scheduling also take place in other types of networks, however, being of significant importance in wireless multi-hop networks due to their decentralized nature and the interference constraints. Regarding the interference, there exist two models defining a successful communication over a wireless channel, the Protocol and the Physical model [81].

Definition 24 *The Protocol Model [81]: Suppose that node i transmits to node j, then node j successfully receives a transmission from i, if:*

$$distance(k,j) \geq (1 + \Delta)distance(i,j)$$

for every node k transmitting simultaneously over the channel,

where Δ is a fixed quantity used when a guard zone is specified by the protocol to prevent neighboring nodes from transmitting simultaneously on the same subchannel. As a result, according to the protocol interference model, there is spatial differentiation among communicating node pairs, meaning that while a transmission takes place, another transmission may occur over the same channel supposing that there is sufficient separation distance between the two. Similarly, more than two transmissions can take place simultaneously supposing that every pair of them is separated by sufficient distance. Therefore, the protocol interference model leads to binary interference constraints, i.e., two transmissions either interfere or not.

For the Physical Model, we first need to make the following assumptions and definitions. Let us suppose that node i transmits directly to node j with power P_i and that the gain over the link ij is $G_{ij} \cong \frac{1}{distance(i,j)^a}$, where $a > 2$ is the path loss exponent (i.e., in this regime the signal power decays with distance). Then if $T(i)$ is the set of nodes transmitting simultaneously with i over the common wireless channel and N is the ambient noise at node j, we define the Signal-to-Interference-plus-Noise-Ratio (SINR) for the transmission of node i to node j as:

$$SINR(i,j) = \frac{\frac{P_i}{distance(i,j)^a}}{N + \sum_{k \in T(i)} \frac{P_k}{distance(k,j)^a}}$$

Definition 25 *The Physical Model [81]: Under the above definitions, the transmission of node i is successfully received by node j, if $SINR(i,j) \geq \beta$, where β is a fixed threshold value determined by the capabilities of the tranceiver devices.*

At this point, we define some metrics used for assessing the performance of wireless multi-hop networks, in the later subsections of this chapter and in other studies as well.

Definition 26 *The Delay (or Latency) of message delivery is defined as the time difference between the generation of a packet from the source node and its arrival at the destination node.*

Definition 27 *Throughput is defined as the number of packets arrived at the destination divided by those sent by the source.*

For the following definition, let us suppose that the network transports one bit-meter when one bit has been transported by one meter towards the destination.

Definition 28 *The Transport capacity [81] is defined as the sum of products of bits and the distances over which they are carried.*

In order to use more compact notation in the sequel, we write that $f(n) = O(g(n))$, if $f(n)$ grows no faster than $g(n)$ and $f(n) = o(g(n))$, if $f(n)$ grows strictly less than $g(n)$. Similarly, we write $f(n) = \Omega(g(n))$, if $f(n)$ grows at least as $g(n)$, and finally we write $f(n) = \Theta(g(n))$, if $f(n)$ grows at the same rate as $g(n)$.

The definitions of this section will hold in the rest of the chapter unless otherwise noted. For an analytical presentation on wireless communications (which is not within the narrow scope of this book), the interested reader can be advised by [81], [156].

6.2   Topology Control (TC) and Inverse Topology Control (iTC)

Multi-hop networks suffer from intermittent connectivity, energy shortage, and wireless medium contention. In order to alleviate these problems, several approaches have been introduced, such as power control [133] and Topology Control (TC) [142]. Contrary to the first, the latter addresses all such issues in a unified, cross-layer framework (oftentimes involving interaction of MAC and Network protocol layers), which is characterized by dynamic and parallel operation. Successful TC protocols are decentralized approaches that vary the transmission power of a node if possible in order to properly adapt its local neighborhood (Figure 6.1). The main objectives of TC are to increase the traffic-carrying capacity of the network by increasing spatial reuse and to reduce energy consumption, while maintaining connectivity and adapting quickly to variations of the topology [96]. Technically speaking, TC may reduce the transmission range of a node if network/node connectivity is not damaged. As a result, the nodes rely on the cooperation with their neighbors for many network tasks such as routing, which can be achieved in a multi-hop fashion. TC attempts to balance a trade-off between energy consumption and node interference on one hand, and network connectivity, on the other. Therefore, a network able to perform TC achieves an exploitation of the node cooperation so as to minimize energy consumption and interference. An additional characteristic of TC protocols is that most of them bear a very efficient and computationally light operation that is appropriate for multi-hop networks, such as the protocols described in [31] and [161]. However, the implementation of a TC protocol on the network topology may lead to increases of the average path length, and deteriorate other performance metrics (except from energy consumption) such as delay and sometimes throughput, which are however critical in modern communication services, especially as the energy capabilities of the nodes are improving. The topology modification mechanism proposed in Section 6.5, which is the main focus of this chapter, employs an inverse Topology Control (iTC) viewpoint. The iTC scheme aims at the opposite result of traditional TC mechanisms, i.e., to improve QoS-relevant performance metrics, e.g., average path length and throughput, while not significantly impacting resource consumption. In other words, starting from a "non-dense" network topology, iTC will rearrange and/or increase/enhance the connections and potentially the positions of special nodes so as to improve the network performance in terms of delay, throughput, path length, and all that with minimum additional cost in terms of energy consumption. In the following, some basic related bibliography of topology modification mechanisms for path length decrease and performance improvements is initially presented (Section 6.4). These approaches resemble the iTC approach. However, most of these approaches lack the optimization and/or the analytical/numerical computation of the additional cost in energy consumption, as it will be explained in Section 6.4. Figure 6.1 shows a possible topology modification, either by the TC or iTC mechanisms, for node A. Topology control could have reduced the radius of node A from value R_f to R'_f, since connectivity is maintained, while energy expenditure at one transmission is reduced due to the range reduction and the same happens for interference. However, after the TC range adaptation to radius R'_f, node A can no longer communicate directly with nodes 2 and 3, but rather only in a two-hop manner through node 5. We do not show the transmission range of nodes 2 and 3 so as to avoid making the figure too complex, but we suppose that these cover node A. This may increase the number of transmissions and retransmissions (increasing in this way the energy consumed totally in the network), the delay, throughput, etc. On the contrary, iTC could have increased the radius R'_f to R''_f so as to achieve a balance between cost metrics (such as energy consumption and interference) and performance metrics (such as delay, throughput, path length), or differently expressed, between the TC approach (range R'_f) and the superfluous connectivity (range R_f).

6.3   Spatial Graphs and Small-World Phenomenon

This section aims at better clarifying the differences between wireless multihop and social network graphs and discusses how these incompatibilities can be overcome in the effort to exploit properties of social networks for improving wireless networks. To begin with, the basic difference between the two graph types lies in the way their nodes are linked together. Wireless multihop networks are spatial graphs, meaning that the possible connections formed between node pairs are restricted by the physical distance separating every two nodes, where an upper bound for the former is in turn determined by the

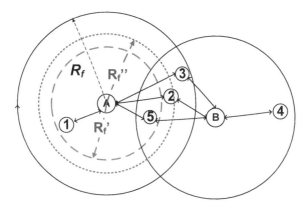

Figure 6.1: Topology Control (TC) vs. inverse Topology Control (iTC). Topology control can reduce the radius of A from R_f to R'_f, since connectivity is maintained and energy cost is reduced. However, then node A can no longer communicate directly with nodes 2 and 3, but rather only in a two-hop way through node 5. iTC can increase the radius R'_f to R''_f so as to achieve a balance between cost metrics and performance metrics, leaning more towards the performance side of the trade-off.

nodes' transmission powers. On the contrary, social networks are relational graphs, i.e., the possibility of connectivity for every pair of nodes in the network is non-zero, irrespectively of the actual distance between them. Any two nodes of relational social networks (e.g., Facebook) can be neighbors.

The $\beta-model$, described in Chapter 5, concerns the case of relational graphs. The construction of relational graphs does not depend on any external metric of distance between the nodes. Therefore, in relational graphs, distances are not taken into consideration in the formation of links. On the other hand, spatial graphs are embedded in a suitable metric space (it is usually the two-dimensional Euclidean space, but it could also be the hyperbolic or any other metric space) and the possible connections between their entities (nodes) depend on their distances defined in the specific embedding metric space. Consequently, if any links are added in a spatial graph, these must conform to the corresponding spatial constraint. Therefore, in order to mimic the $\beta-model$ to add shortcuts in a spatial graph, we have to relax the spatial constraint, so that the desirable length of the shortcut is permitted. Consider that there is a parameter R that defines the maximum distance in which a node can form connections in a spatial wireless network graph. The process of shortcut construction according to the $\beta-model$ consists of the connection of nodes that are otherwise far apart with a particular probability. As a result, in a wireless spatial graph the formation of a shortcut between two nodes (not directly connected in the current graph instance) assumes the relaxation of their power constraints in some way leading to the increase of R for these two nodes. By increasing R, more and longer shortcuts can be formed and when R spans the whole network, every node can be potentially connected with everyone else as in the case of a random or relational graph.

According to [162], spatial graphs can never be totally small-world graphs following the strict definition of Section 5.2.3, and under the naive mechanism of forming longer shortcuts by increasing the transmission radius R, for two main reasons, which are presented and studied in detail in [162]. The first reason concerns the phase transitions of the characteristic path length and the clustering coefficient as functions of R to their random limits (values achieved when R spans the whole network and therefore the spatial graph becomes relational). Contrary to the small-world networks, both phase transitions seem to occur at approximately the same time, or in other words the characteristic path length and the clustering coefficient have similar functional form with respect to R. This happens because the shortcuts for low values of R are not long enough to drastically reduce the path length. When R is increasing, the path length reduces, but so does the clustering coefficient, since the locality of connections is not preserved any more. The second reason is that in contrast to relational small-world graphs, in spatial graphs the path length scaling with respect to the number of nodes n as $n \to \infty$ is not logarithmic. To elaborate more on the first reason, Figure 6.2 illustrates the scaling of the

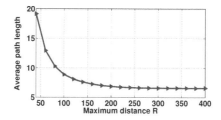

(a) Scaling of the average path length with respect to R.

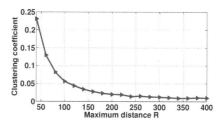

(b) Scaling of the clustering coefficient with respect to R.

Figure 6.2: Scaling of the average path length and the clustering coefficient with respect to R for a spatial graph. We observe that the phase transition of both parameters to their corresponding random limits occurs at the same value $R \cong 150$.

average path length and the clustering coefficient as R increases. Let us denote as k the expected average node degree. The methodology of constructing the undirected spatial graph consists of the following procedure, which is repeated until $\frac{kn}{2}$ links are formed (so that the expected average degree equals to k, since $k = \frac{kn}{2} \frac{2}{n}$) [162]. For each vertex i a node j lying in Euclidean distance at most R from i is chosen uniformly and randomly and the undirected link (i, j) is formed, if it does not exist. The parameters for Figure 6.2 are $n = 1000$, $R = 20 : 400m$, $k = 10$.

However, as we will discuss later, shortcuts in spatial graphs, and especially in wireless multi-hop networks can eventually be realized and in a nominal fashion. Although the resulting graphs are not provably small-world in the strict sense of satisfying the small-world logarithmic length scaling of properties as $n \to \infty$, they are shown, either through analysis or simulations, to develop and exhibit features of small-world graphs and present salient performance improvement. By using specific techniques going beyond the simple technique of relaxing the spatial constraint and adopting features from evolutionary deployment, the average path length and clustering coefficient scaling of the small-world graphs can be approximated closely enough (Section 6.5).

6.4 Inverse Topology Control-Based Approaches

In wireless multi-hop networks the communication type between a node pair depends on their physical distance (for the moment not considering fading) and can be of two types, direct or multi-hop. Nodes being close "enough" can communicate directly, while long-distance nodes use their direct neighbors as relays to achieve a multi-hop communication. This locality of the direct connections, which can be represented as links on a network graph, renders the wireless multi-hop networks highly clustered, almost as clustered as regular graphs. It is observed that multi-hop communications, although they can sometimes increase throughput due to spatial reuse, lead to significant network overhead and delay increment due to the multiple number of physical transmissions taking place for just one packet transfer. Indeed, due to the low reliability of the wireless connections, increase in the number of hops hinders the satisfaction of the necessary Quality of Service (QoS) requirements such as the low total message transmission time (delay) and the high end-to-end delivery rate (throughput). In recent years, some topology modification mechanisms have been developed in order to modify in a realistic way the topology of a wireless multi-hop network (ad hoc or sensor) so as to overcome the aforementioned problems. These mechanisms are inspired by small-world complex networks such as social networks and apply methodologies either centralized or decentralized on the wireless multi-hop topology targeting to reduce its average path length. The common part of these topology modification mechanisms is the enhancement of the wireless network topology with shortcuts in a similar way as proposed in the Watts Strogatz or the Kleinberg's Model (Chapter 5), but adapted to the restrictions imposed by the wireless spatial graphs. We can define two basic categories of such mechanisms with respect to the kind of shortcuts they use, namely wired or wireless.

6.4.1   Early Approaches Using Wired Shortcuts

In this subsection, we will describe some early topology modification mechanisms that augment the network with wires in specific positions aiming to enhance the network topology with small-world features and reduce its average path length. Specifically they deploy some wired connections into the network, with high power transceivers at their ends and are considered to bear no energy constraints, infinitesimal cost of communication, and infinite bandwidth. Since the cost of sending packets along the wires is considered negligible, the communication through the wires is preferable. As a result, wired communications are interpolated among wireless communications, leading eventually to less overhead, delay decrease, throughput increase, and decrease of energy consumption. Wired link placement has both disadvantages and advantages regarding its cost of deployment and performance (especially energy consumption) correspondingly. We will refer more analytically to these aspects after presenting the mechanisms and thus having an idea of how a wireless multi-hop network can be augmented with wired connections. The most important impact of adding wires to a wireless multi-hop network is that it is transformed from a pure wireless network to a hybrid—both wired and wireless—network. Such transformation is not always applicable due to changing the original scope and operation of the network, while this is sometimes incompatible with the operational, environmental, or other constraints/objectives of the network. In the following, we present three of the available topology modification mechanisms for wireless multi-hop and sensor networks, in order to provide a more specific picture of possible integrations of the wired infrastructure in their topologies and the corresponding gains in performance.

Example 1: Randomly placed wires in a wireless grid topology

It is considered a wireless multi-hop network with square grid topology, the topology of which is enhanced via wires' placement aiming to increace its capacity. Such an approach can be found in [137]. Motivated by the random shortcuts of the small-world model of Watts and Strogatz [163], a wired infrastructure consisting of point-to-point randomly placed wired links is chosen. Of course, the final topology is hybrid, including both wireless and wired connections. Regarding the routing procedure, only one wire/shortcut can be used per flow destination pair at each packet transmission and only if it eventually reduces the point-to-point distance in terms of hopcount between the specific source-destination pair. Otherwise, the regular wireless links of the square grid are applicable. The capacity scaling laws of the heterogeneous network are studied in [137] and compared with the case of a pure wireless network and the case where the wired connections are not random but rather form an interconnected square grid. Through these comparisons, it is shown that the random wire placement along with the described communication scheme increases the transport capacity gains in both cases.

In the same fashion, the possibility of improving the energy efficiency and performance (throughput, delay) of a sensor network through its enhancement with wired connections is considered ([47], [145]). The difference between the case of ad hoc networks and sensor networks regarding the topology modification mechanisms and the placement of the shortcuts is the communication pattern. In wireless ad hoc networks we consider the communication taking place between randomly chosen node-pairs, while in sensor networks there are one or more sink nodes and the communication involves the pair of a random sensor and the sink or one of the sinks. The information either begins or it is destined to the sink node(s). As a result, the implemented shortcuts should be oriented to support the communication with the sink, meaning that their placement should be more deterministic and intelligent, leaving its random nature in the case of wireless ad hoc networks.

Example 2: Deployment of wires in a static wireless sensor network with one sink

The enhancement of a wireless sensor network with a single sink node with wires, towards reducing its average path length and decreasing the energy consumption, is investigated [47]. A disk shaped topology where the static nodes are uniformly distributed, and transmit with the same power low rate packets to the also-static sink, arbitrarily located inside the topology, is assumed. The sensors are considered to have local information about the network topology and the locations of wires. Regarding the wires, they have no bandwidth limitations and their ends are equipped with transceivers without energy constraints, which share information with sensors k-hops away from them (local information). The energy consumption for 1 bit of information transmitted from the source to the destination is assumed to be equal to the number of hops of the shortest path separating the source–destination pair [47] (MAC layer issues are ignored), and under this assumption, their aim is to improve the energy efficiency or equivalently the reduction of the average path length. The wire placement is deterministic and it is in accordance with the following procedure: one end of each wire is placed one hop away from the sink node, while the other lies on a circle of radius l, where l is the length of the wire. All the wires added in the network have the same length and are placed on the circle equidistantly, as shown in Figure 6.3. The routing model when there are no wires in the network is greedy geographic routing [80], while when the network is augmented with wires, the source computes the shortest path to the sink, through all the wires for which it has information and also, through only wireless connections choosing the shortest path among them. To make this more clear, if A is the source, B is the sink, and w_{ie1}, w_{ie2} (w_{ie2} closer to B) the transceivers of the wire i, the source computes the distances $dAw_{ie1} + dBw_{ie2}$ (where d expresses Euclidean Distance and it is used for the wireless connections) for all wires i for which A has location information and also computes the distance between A, B, dAB, chooses the shortest path, and forwards the packet to the next hop along this path. It is obvious that the area can be divided into three regions. The first one consists of sensors in a distance less than $\frac{l}{2}$ from the sink that do not use the wires, and the second of sensors in distance higher than $\frac{l}{2}$ from the sink that use the wired connections to send packets to the sink (Figure 6.3). The authors in [47] simulated the proposed topology design in both cases of the sink node placed at the center of the topology and at the end of the topology and for different values of parameters such as the number and length of the wires. The maximum reduction in the path length reaches 70% for the central sink and 60% for the sink at the end. After this percentage, the reduction of the average path length saturates with respect to the wire length. Regarding the restriction of the information up to k-hops away from the wire, when the length of the wire is small, it is not an obstacle in the average path length reduction. However, when the length of the wire increases, the achieved gain in path length deteriorates, especially for small values of k.

Example 3: Combinations of wired shortcuts deployment with routing protocols for a wireless sensor topology (static or mobile)

In the same spirit and towards the reduction of the average path length and the energy dissipation of the sensor nodes, Sharma and Mazumdar in [145], [146] propose four combinations of topology modification and routing schemes based on the addition of wired links. For each created sensor network topology and type of network (static or mobile), a different routing protocol is chosen as more suitable on a per-case basis. The final topology is characterized by basic features of the small-world topology as clustering and small diameter. To begin with the system model, the n sensor nodes are distributed uniformly in a square area with a sink node located arbitrarily in the area. The time is considered slotted, all sensors generate data with the same rate, and transmit by employing the same power level. The network is divided into $\mathcal{N}(n)$ square areas denoted as "cells," each one with side $a(n) = \sqrt{\frac{32 \log n}{n}}$, for mathematical analysis purposes so that a square grid topology is embedded over the network region. In contrast to Example 2 [47], which studies a particular model of wire placement, the main aim here is, given a cost budget for the wires' placement, $l(n)$ ($l(n) < 1$) wires per cell, where n is the number of nodes, to place the wires in such a manner so as to optimally reduce the average hop count in the

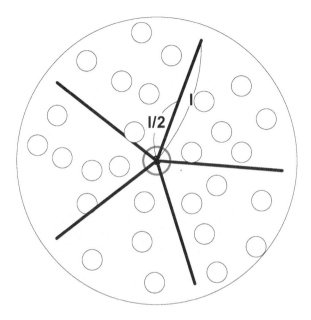

Figure 6.3: A sensor network in a disc deployment area with center at the sink node, enhanced with equidistant wires of equal length, as in [47].

network and achieve the maximum possible gain in energy dissipation. Considering the order of $l(n)$, it is stated that in order to achieve non-negligible gains, $l(n)$ should be of order $\omega(\frac{\log(n)}{n})$. The physical interference model [81] is used, assuming that the network does not change during a time slot. Sensors in adjacent cells can communicate directly, while for indirect communication, the packet is greedily routed to the adjacent cell closest to the sink node, first horizontally up to the column containing the sink node and then vertically towards the sink node. It is supposed that the sensors have knowledge of the cell where the sink node is located. Under this model, the authors prove [145] that each cell almost surely contains $\Theta(\log(n))$ nodes and there is a transmission schedule according to which each cell can successfully transmit once within a fixed number of hops. The last proposition along with the greedy geographic routing guarantee the delivery of packets to the sink node. Similarly to the previous example, the infrastructural support consists of wires equipped with wireless transceivers. Let us define the "Energy Dissipation Skew" (EDS) as the ratio of the maximum to the minimum rate at which a node drains its energy for the transmission and reception of packets, where the maximization or the minimization both span all network nodes. If $EDS = 1$ there is homogeneity in the energy expenditure of the sensor nodes. As the authors prove in [145] and as can also be intuitively considered, in order to minimize the average hop count, the wires should originate from the cell of the sink node. Also, in the same reference it is proved that:

Proposition 1 *Under any wiring and routing scheme that uses at most $\mathcal{N}(n) \times l(n)$ wires the average hop count must be $\Omega\left(\frac{1}{\sqrt{l(n)}}\right)$ and if the path length is $o\left(\sqrt{\frac{n}{\log(n)}}\right)$ the EDS must be $\Omega\left(\frac{1}{l(n)}\right)$.*

In the case of a static sink node, the authors in [145] propose two deterministic schemes (combinations of wired link placement and routing) conforming to the above propositions for adding wired links and for each one of these schemes, they provide bounds and approximations for the EDS and the average hop count. Regarding the first one (scheme 1), every $\frac{1}{l(n)}$ cells are clustered together and the cell in the middle of the cluster is wired to the sink's cell. Supposing that each cell knows the nearest cell with a wire connection, it routes its packet to this cell and from there the packet is directly sent to the sink through the wired connection. In the case of mobile sensor nodes or node failures, the authors propose a protocol where the transceivers on the wire ends periodically transmit their location through Advertisement (ADV) packets and this information is retransmitted by the neighboring cells and so forth, up to a maximum number of hops. Scheme 1 achieves an average path length of order $\Theta\left(\frac{1}{\sqrt{l(n)}}\right)$ and EDS of order $\Theta\left(\frac{1}{l(n)}\right)$. Assuming now that the sensor does not know the exact position of the closest wire and simply (reducing in this way the overhead) forwards with greedy routing the packets to the known cell of the sink node, the achieved path length increases and scales at least as $\Omega\left(\frac{1}{l(n)}\right)$. In this case the gains in path length and EDS of the wire placement are negligible compared with the topology of the pure sensor network. Scheme 2 solves this issue by using a different wires' placement. More precisely, scheme 2 consists of a deterministic placement of less than $\mathcal{N}(n) \times l(n)$ wires and a greedy geographic routing to the sink node, achieving path length and EDS both of order $\Theta\left(\frac{1}{l(n)}\right)$. According to this scheme the wired links placement is performed as follows: if we use index i for the rows and j for the columns of the grid, then all the cells of each row or column of the grid where $i, j = 0 \mod \left(\frac{3}{l(n)}\right)$ are connected with a wired link to the sink node.

Correspondingly, in the case of a mobile sink, the authors propose another two schemes. However, the EDS in this occasion cannot be controlled, and therefore they try to minimize the average hop count and the routing overhead. Also, in contrast to the static sink case and Example 2 [47] the wires cannot all originate from the sink node, since its location is varied. On the contrary, the wire placement is permanent. In the first proposed scheme (scheme 3), the network is clustered as in scheme 1, an arbitrary cluster c_0 is chosen and the cell in the middle of this cluster is deterministically connected with wires with the centers of all the other clusters. A node first sends its packet to the nearest cluster having a wire to c_0, and from c_0 the packet is routed to the sink. This combination of wire placement and routing achieves an average path length of order $\Theta\left(\frac{1}{\sqrt{l(n)}}\right)$. Considering the routing overhead, not only the wire ends should periodically transmit their location, but the mobile sink should also send HELLO messages to make its position public under a total overall overhead of $\Theta(n)$ for each update. This is due to the fact that all the cells should know the nearest wire, and the wire transceivers should know the way to the sink node. The last scheme (scheme 4) proposes a probabilistic wire placement scheme closer in philosophy with the Watts and Strogatz mechanism, where $l(n)$ wires are placed on average per cell and two cells are randomly chosen to become connected with probability proportional to $l(n)$ and a k-power of the inverse of the distance of their centers. Greedy geographic routing is used to send data to the sink node. Scheme 4 achieves an average path length close to $\Theta\left(\frac{1}{l(n)}\right)$ by using greedy routing due to the probabilistic wire placement. In this scheme, the overhead depends only on the mobility of the sink node (due to the greedy geographic routing), and it can be observed that it is capable of leading to an improvement in the overhead with respect to scheme 3 in situations of higher mobility of the sensor nodes compared with the sink node. For more details, the interested reader should refer to [145], [146].

Advantages and Disadvantages of Using Wired Shortcuts

In this subsection, we enumerate some of the advantages and disadvantages [94] of employing wired shortcuts in wireless sensor/ad hoc networks. To begin with the advantages, the wires ensure low-cost transportation of a significant amount of data, as the wired communication is costless with respect to the data transfer compared with the wireless. This is of great importance in a wireless multi-hop network where most of energy is consumed when a sensor transmits and receives packets via the wireless interface. This energy efficiency also contributes to the fact that the transmissions through wires do not interfere with other simultaneous wireless communications, so throughput can be increased overall. In addition, as the wires are shortcuts with low or even zero transmission cost, they can further decrease energy consumption as they reduce the average path length of the network and as a result they also reduce the total number of wireless transmissions in the network. Finally, a wired infrastructure in a wireless multi-hop network can make the network more resilient, keeping the rest of the network interconnected in case of some random node failures.

However, there are a lot of disadvantages in this implementation as well. Firstly, as mentioned above, the network loses its pure wireless character and becomes hybrid, consisting of both wired and wireless network parts. This fact combined with the fact that the wired infrastructure has a costly deployment, especially for short-term applications or non-human-approachable areas (remote surveillance) shows that wires can eventually be deployed only in a small number of applications. Indeed, the wires should be placed offline and there is no possibility of adaptive and on-demand implementation during the network operation. Their deployment has to be designed in a centralized manner prior to the network function. In addition, wired links are in most cases subject to unexpected risks, such as natural disasters and man-made constructions, which can destroy their deployment structure and functionality. Finally, they do not fit well to sensor networks with mobile sink nodes, or in general wireless network cases with mobile nodes.

From all the above, it can be concluded that although the wired shortcuts have important advantages, especially regarding the energy consumption of the wireless network, their disadvantages eventually prohibit rendering the deployment of wired shortcuts unrealistic in the majority of wireless networks. In the next section, we study alternative approaches regarding topology enhancement with wireless shortcuts that seem to be more promising for a possible realistic and tangible implementation of a "small-world" wireless multi-hop network. Most importantly, when the added shortcuts are wireless, the network can always return back to its energy-conserving mode, where the nodes transmit in lower power, when there is no need of such added higher power, long-range transmissions, as the ones added for the QoS performance improvements.

6.4.2   Approaches Using Wireless Shortcuts

Helmy in his work in [86] was the first to study the possibility of inducing small-world properties in the spatial graphs of wireless multihop networks by using wireless shortcuts while [86] does not provide a specific framework for enhancing a wireless multi-hop network with wireless shortcuts, and this possibility is mostly explored through simulations. Also investigated was what advantages can be derived by such an enhancement especially in the field of developing efficient protocols for large scale wireless networks (ad hoc and sensor), i.e., resource discovery. The wireless shortcuts may represent either physical links or logical links that translate into multiple physical hops. The wireless multihop network is represented initially by a Random Geometric Graph (RGG). In order to study the benefits from the enhancement of a small-world graph with features of social networks, two simple topology modification mechanisms are used: link addition and link rewiring. In link rewiring, a node is chosen at random and a link to one of its neighbors is removed and relinked to another randomly chosen node (with proper power adaptation of the participating nodes if needed). In link addition, a link is added between a randomly selected pair of nodes in the network (also with proper power adaptations as in link rewiring). Two main results were derived from Helmy's study. The first one deals with the small-world character of the yielded network topology after link rewiring. Specifically, the derived topology has small-world features, since by rewiring a small fraction of the links (0.2%–2%) the average path length is drastically reduced, while at the same time the structure of the network, expressed by the clustering coefficient, is not damaged. The second conclusion regards the possible increase in the transmission overhead following the addition of shortcut edges. However, the simulation results indicated that by limiting the shortcut length to a fraction of the network diameter (and more specifically to $\frac{D}{4}$, where D is the network diameter) the average path length is maximally reduced and therefore it can be concluded that the length of the shortcuts, which is strongly linked with the additional interference, can be limited. In the same spirit as [86], in [4] is proposed a simple, optimized placement of long-range links in a square grid, wireless mesh network. The main aims of the optimization approach of shortcut placement is to reduce the delay due to multi-hop transmission and increase capacity (throughput), while taking into consideration practical aspects such as interference and traffic congestion imposed by the added shortcut links. The equal-length, long-range links are established between randomly selected pairs of nodes, which are assumed to be equipped with directional antennas for the targeted long-range communication and two radio transceivers (one used for the short and the other for the long connection). Only these selected nodes transmit in higher power. As expected, the random placement of shortcuts in the square grid leads to small-world features. However, a more deterministic shortcut-topology is achieved through genetic algorithm optimization (Chapter 3), which is characterized by minimizing heuristically the average path length. Similar heuristic optimization methods with respect to shortcut placement are also adopted aiming at reducing interference and achieving load balancing. Finally, through simulations it is concluded that the number of shortcuts employed should be of the order of the initial average path length of the network.

Up to now, the described mechanisms have not been suitable for a realistic implementation as they do not take into consideration important limitations regarding the nature of wireless multi-hop networks. Firstly, the wireless links have a limited bandwidth, contrary to the wired ones, and therefore there is a limit in the information carried through them. Also, each node may have a limited number of separate frequencies K_{LL} and an upper bound on its transmission power (or transmission radius) R_h, posing a limit to the potential maximum length of the long-range link. In [158], the authors take all the above constraints into consideration, and propose three mechanisms for reducing the average path length of a wireless mesh network with a central gateway. The mesh network consists of static routers that provide services to wireless clients through multi-hop transmissions up to the also-static gateway node (Figure 6.4). Each node, excluding the gateway, has one short-range radio and K_{LL} long-range radios for establishing shortcuts with distant nodes. The gateway is assumed to form only short-range links. In all the mechanisms, the shortcuts added must be bidirectional, the Euclidean distance between the selected nodes must be less than R_h, while both selected nodes must have a spare long-range radio, since each node assigns a separate radio to each one of its long-range connections. The first mechanism, simply places shortcuts between selected pairs of nodes satisfying the above constraints. In

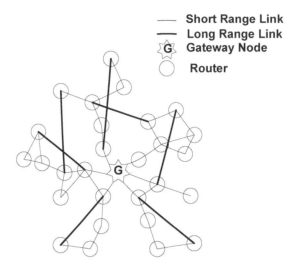

Figure 6.4: Topology enhancement with long range links as described in [158].

the second mechanism, an additional constraint is posed for the difference of the distances of the two endpoints (nodes) of the possible shortcut from the gateway node, denoted as Δ_h, to exceed a certain value, Δ'_h. However, the third mechanism is proved by simulations to be the most efficient, according to which the pairs of nodes with higher Δ_h are first greedily linked giving rise to higher reductions of path length toward the gateway node. For all mechanisms, the path length toward the gateway is reduced, while this reduction increases with the increase of the number of shortcuts in the network and the values of R_h, K_{LL}, Δ'_h. However, for the first two parameters, saturation points are presented through simulations, as in [47], after which the path length cannot be reduced more. Simulations show that the long-range connections are not congested in contrast to the common expectations and in fact assist the network to serve more load than the initial traditional mesh network.

The previous approaches have a common drawback, which consists of the fact that the shortcuts are created irrespectively of the communication needs and their length is predetermined. Also, the nodes need to be stationary as the shortcut placement enhances a particular network topology. The authors in [94] follow a completely different approach, where the shortcut creation and its length are determined by the data demands. Another important contribution of [94] is the extension of these mechanisms on disconnected networks in contrast with the rest of the related works, which refer to connected networks. Therefore, the mechanism proposed in [94] applies on connected and partially-connected (delay-tolerant) ad hoc and sensor networks. More explicitly, instead of increasing the radios' transmission ranges or enhancing the network with wired shortcuts, a small number of mobile nodes denoted as "data mules" is used to mimic shortcuts in the wireless network. These data mules also forward packets between disconnected parts of the networks. According to the proposed mechanism, the normal nodes execute greedy forwarding routing, when they have packets to send. However, when there is a neighboring data mule moving toward the direction of the destination, the data packet, instead of being greedily routed to the next hop normal neighbor, is attached to the data mule. If the data mule reaches the destination, or starts moving in the opposite direction of the destination, data is unloaded. Indeed, if the data mule turns, it is necessary to check whether its direction remains toward the destination for continuing to carry the data. The simulation results indicate that a small number of data mules ($\sim 10\%$ of the total number of nodes) reduces the average path length of a connected wireless network up to 50%. However, the data mules increase the delay of message delivery, so the emerging trade-off must settle in an equilibrium state. In disconnected networks, employing a number of data mules up to approximately 30% of the total number of nodes increases the packet delivery rate, and enhances the network with small-world features. Finally, it should be mentioned that if the initial network contains some mobile nodes, these can act as data mules, and as a result, there is no need for extending the network with new mobile nodes (delay-tolerant case).

A completely different approach, using aerial nodes, is presented in [15], where the authors focus on the topological properties of the network in order to improve the performance (i.e. the convergence) of distributed algorithms. In this direction they create a small-world topology aiming to ensure a faster convergence of distributed algorithms in the whole network. More specifically, they aim at creating topologies that are simultaneously sparse (i.e., each node is connected to few other nodes) and highly connected. Such efficient topologies are the expander graphs (Section 5.5) and the small-world graphs (Section 5.2), which offer a favorable trade-off between convergence speed and cost of collaboration. In [15], the network is first clustered and centrality-based criteria (locally computed betweenness centrality) are used to choose the cluster heads. The convergence is achieved firstly at every cluster and then in the inter-cluster area among the cluster heads. The convergence further speeds up with the aid of aerial nodes that ensure the connectivity of cluster heads. Although being costly, the connections of the clusters with the aerial nodes function as shortcuts or more correctly as contractions as there are defined in [162], by connecting nodes that are otherwise multiple hops apart. The employed number of the aerial nodes is minimized subject to connectivity maintenance balancing the trade-off between cost of deployment and group performance. Finally, the authors propose an expander graph structure at the level of the aerial nodes so as to speed up their peer communication (see the applications of expander graphs in Section 5.5).

Eventually, a more holistic approach is presented in [147] and [148]. In these works a more general framework is analytically described and mathematically analyzed through continuum theory and simulated in Matlab and Ns2. In practice, the infusion of wireless long-range links in a wireless multi-hop network is based on either enhancing the network with nodes-hubs with increased power capabilities or adding directed long-links between wireless node pairs. This idea is captured in the general framework presented in [147] and [148], which is based on network churn. Network churn consists of two sub-mechanisms, edge churn and node churn, where the first one includes the addition, deletion, and rewiring of links and the second the addition and deletion of nodes. These processes naturally take place in dynamic systems, through the evolution of the wireless multi-hop network and not only as a part of a biased modification mechanism. However, the topology modification mechanism emerging from the above framework properly adapts edge churn and node churn, so as to achieve the infusion of social/small-world properties in the wireless multi-hop network by controlling the trade-off between the cost-budget (energy consumption and radio resources) and the desirable performance. More specifically, the probabilistic framework presented takes place in a limited number of time slots during which the network is considered static and their number is determined by the capabilities of the network, such as the maximum possible transmission power of a node. Therefore, the length of the added shortcuts is determined by the nodes' power capabilities, and neither an excessive transmission power of the nodes, nor the use of different transmission frequencies are required, although the latter implementation can be easily incorporated in the mechanism. Therefore, the proposed framework is rather realistic and fits in the pure wireless network's case paradigm. In addition, a distributed implementation of the framework based on given parameters' values is described, therefore giving the possibility of an independent deployment of increased transmission radii in the network when this is desired from network performance. However, when the network performance is high, the nodes can always return to their original, less-power-consuming modes by reducing their transmission radii. We will present this general mechanism and mathematically tractable special cases of it in complete detail, in the following section.

In [78], a theoretical model of topology modification for a wireless sensor network based on wireless shortcuts is described and simulated. As well as the theoretical model, a distributed protocol was also developed for on-line implementation of the mechanism, during the operation of the network. This proposed distributed implementation is very important for the realization of small-world features in a wireless sensor network in an on-demand, reactive manner. The latter means that the shortcuts can be employed in case the performance of the traditional sensor network deteriorates and needs improvement. Because of the sensor network topology, the placement of shortcuts is oriented towards the sink node. Specifically, there are two types of sensor nodes in the network, the L-sensors with low power capabilities and the H-sensors, which are equipped with more powerful hardware, and are able to transmit over long distances. The shortcuts are unicast links between the H-sensors, which function in two separate frequencies for their communication with other H-sensors or L-sensors correspondingly, aiming to reduce inter-

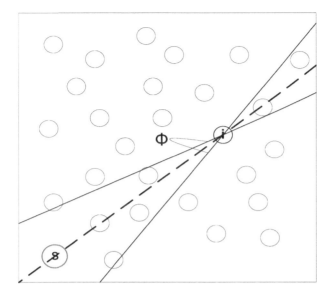

Figure 6.5: Topology of the sensor network. H-sensor i can create a unicast long range link towards the sink with another H-sensor lying inside the marked angular area ϕ.

ference. In the theoretic model, an edge is selected and one of its ends, e.g., i, is linked with probability p with another node lying in a sector of angle ϕ centered on i, the bisector of which contains the sink node, as shown in Figure 6.5. This procedure is followed for all edges of the network graph. In the online protocol, however, this procedure is not efficient as each node has to know the coordinates of all the other sensors in the network. Also, this process leads to a pre-planned topology of H-sensors as we force the H-sensors to be the endpoints of the randomly added shortcuts, which might not be the case for the real network. Therefore, the procedure is adapted, so as to choose only H-sensors as possible endpoints of long-range links and so each H-sensor has to know only the location of other H-sensors in its communication range. In order to achieve this simplification, the theoretical model has to be pre-evaluated through simulations with the specific probability p and a new probability p' should be computed expressing the probability of creating a shortcut between two H-sensors. p' is defined as the fraction of the average number of shortcuts of an H-sensor divided by the average number of H-sensor neighbors of an H-sensor. Experimental evaluation of the online model shows an emerging trade-off between throughput increase (the number of packets transmitted to their destination from the sink) and latency (delay) decrease on the one hand, and the energy consumed in the network or else the number of H-sensors on the other. Increment of p leads to an increment in throughput and latency reduction, but also an increase in the energy consumed from the H-sensors. According to [78], the value of p that creates a useful equilibrium in the above trade-off is $p = 0.01$.

Wireless shortcuts are strongly linked with the increase of the transmission range of the wireless nodes. Therefore, deeper research is needed, since by increasing the transmission radius of the wireless nodes, interference and load balancing issues can deteriorate network performance. This is a weak point of all the mechanisms existing until now for pure wireless networks (and an advantage of the wired shortcuts regime), as it is not proven in a formal sense that the infusion of small-world properties in the network topology can safely improve throughput and delay. Analytical results that prove bounds on the network capacity for the "small-world" wireless ad hoc and sensor networks are available only for certain types of enhanced topologies [107], [137]. Proving the throughput gains of the modified topologies is essential, especially in the case of adding wireless shortcuts in order to create a pure wireless "small-world" network. Contrary to the case of enhancing the topology with costless wired shortcuts in terms of communication, the wireless shortcuts are obviously strongly related to transmission radii increments. This fact, as it is shown in [91], can reduce throughput, due to node activity reduction, if the shortcuts are not appropriately chosen and realized. Beyond the mathematical analysis of the performance, most mechanisms of the above presented [4, 15, 86, 148, 158] do not test their mechanisms in a real environment (i.e. Testbed) to provide validation for possible gains in throughput and delay. Also, it is interesting to further study the interaction between some scheduling/routing schemes and network topology so as to find the best combination of a distributed scheduling/routing and enhanced topology for providing throughput and delay optimality.

6.5   Holistic Topology Modification Framework

In this section, we describe and analyze a holistic topology modification framework for weighted network graphs. Binary graphs constitute a special case and can be taken into account by simply considering all the weights equal to 1. They will be studied in detail in Section 6.6.1. The main scope of this topology modification mechanism is to enhance the network topology with small-world properties in an "online" way, while simultaneously accounting for the particularities of each network, or for the overlaying applications, both expressed through the link weight values. The term "online" expresses the possibility of applying the topology modification mechanism when this is required by the network performance requirements and not "a priori." The holistic framework consists of two basic mechanisms, inspired by dynamic network evolution, the Weighted Edge Churn (WEC) mechanism, which is applied over the network edges, and the Weighted Node Churn (WNC) mechanism, which is applied over the network nodes. Each one of these mechanisms has two components, an addition and a deletion process of the corresponding participating entities, i.e., for the links in WEC and for the nodes in WNC. Contrary to the topology modification mechanisms of the previous section (Section 6.4.2), the holistic topology modification framework is not limited to only the addition of long-range edges and nodes-hubs (edge and node addition processes), but also incorporates deletion processes as well. The intuition behind this enhancement is to both add new edges and nodes in a suitable manner for contracting the path length, and at the same time, delete appropriately selected links and nodes for balancing the cost of various network functions.

First, we describe the system model considered for the application of the proposed topology modification mechanism. The time is considered slotted and the nodes synchronized. A square deployment area of side L is considered, where N nodes with homogeneous transmission radius R_f are randomly and uniformly distributed. Also, it is assumed that an initial weight in a specified interval is assigned to each directed link. Each node is capable of varying its current transmission radius, denoted as $R_c(i)$, in the range $[R_{MIN}, R_{MAX}]$, i.e., there is a restriction on the maximum transmission power of nodes rendering the mechanism suitable for a realistic implementation. Each time step is characterized by two parameters, $R_{\min}(t), R_{\max}(t)$, where $R_{MIN} \leq R_{\min}(t) \leq R_{\max}(t) \leq R_{MAX}$. These two parameters will determine the maximum and minimum possible length of a shortcut at time slot t. Initially, $R_{MIN} = R_{\min}(0) = R_{\max}(0) = R_f$. At the beginning of each step t (starting at $t = 1$), $R_{\min}(t) = R_{\max}(t - 1)$ and $R_{\max}$ is increased by a predefined step value a, i.e., $R_{\max}(t) = R_{\max}(t - 1) + a$. In order to avoid increasing the probability of disconnecting the network topology via the deletion processes, we assume higher addition rates of links and/or nodes than the corresponding deletion rates, so as to maintain connectivity in the final network graph with high probability. Next, the weighted edge and node churn mechanisms are described and analyzed, and afterwards, they are combined in one holistic modification framework for weighted graphs.

6.5.1 Weighted Edge Churn Framework

Weighted Edge Churn (WEC) is the process of dynamic link evolution, consisting of dynamic link addition and link deletion. Link addition, along with the node addition process of WNC, are the main processes of the infusion of small-world features in the topology of wireless multihop networks. Link deletion aims mostly at balancing the trade-off between the path length reduction and the cost incurred by overcrowding the network with links. The endpoints of the new links are selectively chosen according to some probability distribution (Appendix A), which determines with high probability connections to more suitable nodes depending on the overlaying application and the operational objectives/requirements. Due to the existence of a weight value on each link, the new links should be assigned an initial weight value in an appropriate way, while the existing weight values should be re-adapted due to the change of the network topology. To make this clearer, in the case that the

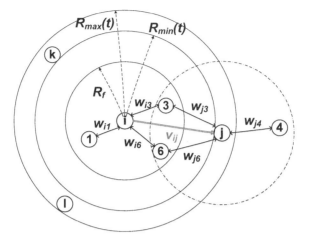

Figure 6.6: Edge addition and weight adaptation.

weight values represent communication data flow, the addition of a new link will force changes on the respective traffic flows. Therefore, the weights should be suitably adapted to the new network topology. Next, WEC is described and analyzed in detail.

Process of Edge Addition

The edge addition process takes place at step t with probability p $(0 \leq p \leq 1)$ and m_a new connections are added to m_a selected nodes. Each one of the m_a nodes is selected with probability $\mathbb{P}_i^a$ in order to add a link and increases its range for including its new one-hop neighbor. Similarly to online social networks, whose members obtain new acquaintances, the m_a selected nodes become more "social" (popular) in the network. The probability $\mathbb{P}_i^a$ is proportional to node i's "popularity," thus proportional to the intensity that the other network nodes want to communicate with i, since popularity in terms of communication denotes traffic intensity. Edge addition is realized at both the Physical and the Network protocol stack layers. More explicitly, if link addition is performed at step t, a node i is selected with probability $\mathbb{P}_i^a$[2] $(\sum_{i=1}^N \mathbb{P}_i^a = 1)$ and extends its range from $R_c(i)$ to $R_{\max}(t)$, as shown in Figure 6.6. Although i increases its transmission range to $R_{\max}(t)$ and thus, it can transmit in the Physical layer to all the nodes in this range, it forms a Network layer "long-range" connection, only with one node j, lying in the annulus from $R_{\min}(t)$ to $R_{\max}(t)$ centered at node i and denoted by $\mathcal{A}_{R_{\min}}^{R_{\max}}(i)$. Thus, node i avoids depleting its energy by sending and forwarding to all the nodes in the newly added range. Node j is selected among all nodes in $\mathcal{A}_{R_{\min}}^{R_{\max}}(i)$

---

[2]The notation $\mathbb{P}_i^a$ denotes the probability of selecting i to add a link. a should not be confused with a numerical exponent of $\mathbb{P}$.

with probability $Q(j|i)$ ($\sum_{j \in \mathcal{A}_{R_{\min}}^{R_{\max}(i)}} Q(j|i) = 1$), which denotes the selection of j given that i has already been selected. The form of $Q(j|i)$ depends on the application and may be a function of the weights in the local neighborhoods of nodes i, j, i.e., $Q(j|i) = f_1(w_{kh}|k, h \in$ local neighborhoods of $i, j)$. Also, the weight v_{ij} of the new link is determined according to the application setting and $v_{ij} = f_2(w_{kh}|k, h \in$ local neighborhoods of $i, j)$. The addition of the directed link $i \to j$ with weight v_{ij} is depicted in Figure 6.6, where initially j was two hops away from i and $R_c(i) = R_f$. At slot t, $R_c(i)$ becomes equal to $R_{\max}(t)$ and j becomes a one-hop neighbor of i. However, node i still communicates in a multi-hop manner with nodes k, l lying in the ring delimited by radii $R_{\min}(t), R_{\max}(t)$ and centered at node i. The above process is repeated for each one of the m_a nodes selected at a time step t and the links added are considered directional, as only the transmission radius of the node initiating the modification process increases. As a result, the final network topology can be represented by a directed network graph, since the new links added are directed, as it will be explained in the sequel. We denote by $\mathcal{N}^{in}(i)$, the set of nodes having a direct connection towards node i and by $\mathcal{N}^{out}(i)$ the set of nodes towards which i is connected. By controlling parameters p and m_a, different combinations of the number and length of the added links can be achieved. More explicitly, we can either add a few links at a step but of different lengths, i.e., in many steps (with high p and low m_a), or add many links of the same length (low p, high m_a), or even add many links at each step of different lengths (high p, high m_a), etc.

The addition of the new weighted link will dictate a corresponding change in the weight values of the rest of the links in the network depending on the metric these weights represent. As an example, let us suppose again that the weight of each link represents its traffic flow. In this case, if a node adds a new connection towards node i, the traffic passing through i will be increased, leading to a corresponding increase of the flows of the links starting from node i, which in the sequel increases the flows of i's one-hop neighbors, and so on and so forth. For simplification, at this mechanism, it is considered that the addition of a long-range link (i, j) will cause a variation in the weights of only the local neighborhood of nodes i and j, similarly to the weight adaptation of Section 5.3.4, as shown in [18] as well. It is assumed that a node is able to adapt only its out-links, which is a reasonable assumption, as the traffic passing through the out-links of a node is controlled totally by this node, while the traffic passing through the in-links is controlled by neighboring nodes. Precisely, as in Figure 6.6, node i has added a long-range connection $i \to j$ and thus it can use more intensively this connection for traffic exchange towards this flow direction, while it can reduce the traffic sent through the rest of its out-connections towards the specific direction. Similarly, node j has obtained a new in-connection that will result in more flow leaving node j, especially when j is used as a relay node. Following a similar approach for the weight adaptation as in [18], [20] described in Section 5.3.4, three constants, δ^{1a}, δ^{La}

and δ^{2a} are defined to express the total change in the amount of traffic, for node i, for the link $i \rightarrow j$ and node j correspondingly. As a result, node i adapts the weights of its out-links as follows:

$$w'_{ih} = w_{ih} + \delta^{1a} \frac{w_{ih}}{s_i^{out}}, h \in (1, 2, .., N), h \neq j \tag{6.1}$$

The link $i \rightarrow j$ is adapted as:

$$v'_{ij} = v_{ij} + \delta^{La} v_{ij} \tag{6.2}$$

Similarly for j,

$$w'_{jh} = w_{jh} + \delta^{2a} \frac{w_{jh}}{s_j^{out}}, h \in (1, 2, .., N) \tag{6.3}$$

After the addition of the directed link $i \rightarrow j$ (Figure 6.6) the strengths of nodes i and j are altered as follows:

$$s_j^{in} \leftarrow s_j^{in} + v_{ij}(1 + \delta^{La})$$
$$s_j^{out} \leftarrow s_j^{out} + \delta^{2a}$$
$$s_i^{out} \leftarrow s_i^{out} + v_{ij}(1 + \delta^{La}) + \delta^{1a} \tag{6.4}$$

At this point, the construction of the differential equations expressing the rate of change of the in-strength and out-strength of each node, is outlined. Towards this direction a Continuum Theoretic approach is utilized and thus, a similar logic as for the construction of the Eq. (5.18) of Chapter 5 is adopted as well. The out-strength of node i changes when:

1. The node is the one initiating the process, i.e., the node from where the long-range link starts (node i in Figure 6.6). In this case, i is selected with probability $\mathbb{P}_i^a$ and s_i^{out} changes by $v_{ik}(1+\delta^{La})+\delta^{1a}$ with k chosen with probability $Q(k|i)$.

2. The node i is the end of the long-range link (i.e., node j in Figure 6.6). This occurs if one of i's neighbors, i.e., $k \in \mathcal{A}_{R_{\min}}^{R_{\max}}(i)$ is chosen to initiate the process and chooses i with probability $Q(i|k)$. In this case s_i^{out} changes by δ^{2a}, and this takes place with probability

$$P_G(i) = \sum_{k \in \mathcal{A}_{R_{\min}}^{R_{\max}}(i)} \mathbb{P}_k^a Q(i|k) \tag{6.5}$$

However v_{ik} does not depend only on i and this creates a difficulty in constructing the differential equation for the rate of change of s_i^{out}. Thus, instead of v_{ik}, the average of the v_{ik} for all neighbors k of i in the $\mathcal{A}_{R_{\min}}^{R_{\max}}(i)$ is defined and used, expressed by equation:

$$\hat{v}_i = \frac{\sum_{k \in \mathcal{A}_{R_{\min}}^{R_{\max}}(i)} v_{ik}}{\frac{N|\mathcal{A}_{R_{\min}}^{R_{\max}}(i)|}{L^2}} \tag{6.6}$$

From all the above, the rate of change for s_i^{out} is:

$$\frac{ds_i^{out}}{dt} = m_a p \left[\mathbb{P}_i^a (\hat{v}_i (1 + \delta^{La}) + \delta^{1a}) + P_G(i)\delta^{2a} \right] \tag{6.7}$$

Similarly, the in-strength of node i changes when:

1. The node is an out-neighbor of the one initiating the process (nodes $1, 3, 6$ in Figure 6.6). This happens with probability $\sum_{k \in \mathcal{N}^{in}(i)} \mathbb{P}_k^a$.

2. The node is the end of the long-range link (node j in Figure 6.6) with probability $P_G(i)$.

3. The node is an out-neighbor of the end of the long-range link (nodes $3, 4, 6$ in Figure 6.6). This happens with probability $\sum_{k \in \mathcal{N}^{in}(i)} P_G(k)$.

Thus, the evolution equation for the s_i^{in} is:

$$\frac{ds_i^{in}}{dt} = m_a p \left[\sum_{k \in \mathcal{N}^{in}(i)} \mathbb{P}_k^a \delta^{1a} \frac{w_{ki}}{s_k^{out}} + \right.$$

$$\left. \sum_{k \in \mathcal{A}_{R_{\min}}^{R_{\max}}(i)} \mathbb{P}_k^a Q(i|k)(\delta^{La} + 1)\hat{v}_k + \sum_{k \in \mathcal{N}^{in}(i)} P_G(k)\delta^{2a} \frac{w_{ki}}{s_k^{out}} \right] \tag{6.8}$$

Process of Edge Deletion

Edge deletion takes place at step t with probability q ($0 \le q \le 1$), and m_d links in total, one from each of m_d selected nodes, are deleted. A node i is selected to initiate edge deletion with probability $\mathbb{P}_i^{d3}$ ($\sum_{i=1}^{N} \mathbb{P}_i^d = 1$), which depends on the popularity of a node. More precisely, less popular nodes are chosen with high probability to perform deletion. This choice follows typical trends in social networks, but it is also intuitive, since deleting one of the out-links of an unpopular node would not significantly influence the informa- tion flow. It is considered that, in the case that both directions of the link between nodes i and j exist, only the link $i \to j$ is deleted. This can be achieved at the Network layer of the Protocol Stack through proper routing modification. The node j from where the edge is deleted is chosen with prob- ability $Q^d(j|i)^4$ ($\sum_{j \in \mathcal{N}^{out}(i)} Q^d(j|i) = 1$) among the out-neighbors of i where $Q^d(j|i) = f_3(w_{kh}|k, h \in \text{local neighborhoods of } i, j)$. As an example, in Figure 6.7 node i is selected to delete its link with its one-hop neighbor j. Similarly to

---

[3] The notation $\mathbb{P}_i^d$ denotes the probability with which i is chosen to initiate the edge deletion process. d should not be confused with a numerical exponent of $\mathbb{P}$.

[4] In this notation d should not be confused with a numerical exponent of Q.

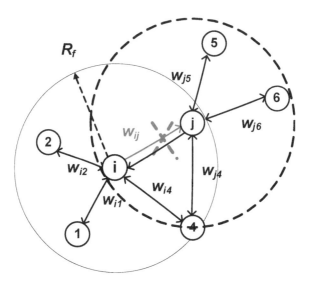

Figure 6.7: Edge deletion and weight adaptation.

edge addition, we consider an adaptation of weights in the local neighborhood of nodes i and j, due to the deletion of the link $i \rightarrow j$. Therefore, the constants δ^{1d}, δ^{2d}, express the total amount of change of flow in the remaining links of i and j respectively and the following weight adaptations take place:

$$w'_{ih} = w_{ih} + \delta^{1d} \frac{w_{ih}}{s_i^{out} - w_{ij}}, h \in (1, 2, .., N), h \neq j \qquad (6.9)$$

Similarly for j, the adaptation of the weights of its out-links are given by:

$$w'_{jh} = w_{jh} + \delta^{2d} \frac{w_{jh}}{s_j^{out}}, h \in (1, 2, .., N) \qquad (6.10)$$

After the deletion of the directed link $i \rightarrow j$ (Figure 6.7) the strengths of nodes i and j are re-computed as follows:

$$s_j^{in} \leftarrow s_j^{in} - w_{ij}$$
$$s_j^{out} \leftarrow s_j^{out} + \delta^{2d}$$
$$s_i^{out} \leftarrow s_i^{out} - w_{ij} + \delta^{1d} \qquad (6.11)$$

Sequentially, following the same approach as in the edge addition mechanism, the rates of change of the strength values are computed through differential equations [20]. Specifically, the out-strength of node i changes when:

1. The node is the one initiating the process, i.e., the node deleting an out-link, e.g., $i \rightarrow j$ (node i in Figure 6.7). In this case, node i is selected with probability $\mathbb{P}_i^d$ and s_i^{out} changes by $\delta^{1d} - w_{ij}$.

2. The node i is losing an in-link (i.e., node j in Figure 6.7). This happens if one of its neighbors k in $\mathcal{N}^{in}(i)$ is chosen to initiate the process and chooses i with $Q^d(i|k)$. Thus the probability is

$$P_G^d(i) = \sum_{k \in \mathcal{N}^{in}(i)} \mathbb{P}_k^d Q^d(i|k) \qquad (6.12)$$

In this case s_i^{out} changes by δ^{2d}.

However w_{ij} does not depend only on i and this creates a difficulty in constructing the differential equation for the rate of change of s_i^{out}. In a similar way that the similar problem of the weight of the new link, v, is handled in the edge addition process, instead of using w_{ij}, we use the average weight, $\bar{w}$, of the links of the whole network.

Consequently, the equation for the evolution of s_i^{out} is:

$$\frac{ds_i^{out}}{dt} = m_d q[\mathbb{P}_i^d(\delta^{1d} - \bar{w}) + P_G^d(i)\delta^{2d}] \qquad (6.13)$$

In a similar manner with the description of the changes in the out-strength, the in-strength of a node i changes when:

1. The node is an out-neighbor of the one initiating the process (nodes $1, 2, 4$ in Figure 6.7). This could happen with probability $\sum_{k \in \mathcal{N}^{in}(i)} \mathbb{P}_k^d$.

2. The node is the node from where the in-link is deleted (node j in Figure 6.7) with probability $P_G^d(i)$.

3. The node is an out-neighbor of the node that loses its in-link (nodes $4, 5, 6$ in Figure 6.7), which happens with probability $\sum_{k \in \mathcal{N}^{in}(i)} P_G^d(k)$.

Thus, the equation describing the evolution for the s_i^{in} is:

$$\frac{ds_i^{in}}{dt} = m_d q\left[\sum_{k \in \mathcal{N}^{in}(i)} \mathbb{P}_k^d \delta^{1d} \frac{w_{ki}}{s_k^{out} - \bar{w}} - P_G^d(i)\bar{w} + \sum_{k \in \mathcal{N}^{in}(i)} P_G^d(k)\delta^{2d} \frac{w_{ki}}{s_k^{out}} \right]$$

$$(6.14)$$

Weighted Edge Churn (Edge Addition and Deletion)

After analytically describing the processes of link addition and link deletion, the next step is to unify them in a common framework so as to balance the trade-off between increased cost due to the addition of shortcuts and performance (or average path length) improvement. Combining edge addition and deletion is important, since the first infuses small-world features in the multi-hop topology (leaning towards performance), while the second ensures that the induced network does not become overcrowded and thus it does not suffer from excessive interference (restricting the respective entailed cost). Summing up, in order to formulate the unified framework, in WEC mechanism, at time slot t, edge addition takes place with probability p and m_a links are added to m_a appropriately selected nodes, while edge deletion takes place with probability q and m_d selectively chosen links are deleted. We assume that $p + q \leq 1$ and if $p + q < 1$ then with probability $1 - p - q$ the network topology does not change. In the combined edge addition/deletion mechanism, the rate of change $\frac{ds_i^{in}}{dt}$ is obtained by the superposition of the rates in Eq. (6.8) and (6.14), while the rate $\frac{ds_i^{out}}{dt}$ is obtained by the superposition of the rates in Eq. (6.7) and (6.13). However, in the general case, the solution of these equations is cumbersome and requires that all quantities included (weights, probabilities) are expressed as time functions. In the following section (Section 6.6), the WEC mechanism will be restricted to the case of binary graphs (using node degree instead of node strength) and analytical solutions will be obtained for the rate of change of the node degree. The average change of the number of links at each time step t due to edge churn is equal to $pm_a - qm_d$. Thus, up to step t, we have a total average change of $\ell(t) = (pm_a - qm_d)t$ of the number of links in the network.

6.5.2   Weighted Node Churn Framework

Based on and extending the Weighted Edge Churn Framework, the Weighted Node Churn Framework is developed, which applies addition and deletion of nodes in weighted topology graphs. Node variation as the basis of Weighted Node Churn is a natural process in many wireless multi-hop networks. For instance, some nodes may deplete their energy and become inactive, while others recharge and become operational again. Node churn is also evident in online social networks where new users enter the network, possibly invited by their friends, or existing nodes choose to leave the network especially when their intensity of participation in the network processes is low. Therefore, inspired by social networks, the addition and deletion processes of the node churn mechanism can be designed in such a way that the spatial wireless multi-hop network graph exhibits features encountered in small-world networks.[5]

Process of Node Addition

At time slot t, node addition takes place with probability r and M_a new nodes are added in the network. The new nodes have increased radius compared to the original ones (greater than R_f) in the network, which serves the purpose of introducing in the network "hub" nodes for eventually reducing the average path length as in the case of scale-free networks (Section 5.3). Each new node, added at step t, has a radius $R_{\max} = R_f + at$. The new connections are considered bidirectional in an area of πR_f^2 and directional in the annulus extending from radius R_f to radius $R_{\max}$ centered at the newly added node

---

[5]We call these features small-world-like features.

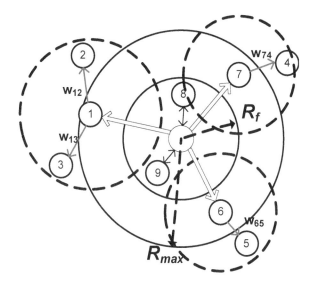

Figure 6.8: Node addition and weight adaptation.

as shown in Figure 6.8. The weights of the new links are chosen according to a probability distribution (Appendix A) depending on the application. For instance, the beta $\beta(m, n)$ distribution can be used, as it takes different forms depending on m, n and thus can be adapted to a wide range of applications.

The addition of the new node with increased radius is considered to cause a variation in links' weights explained in the following. Inside the disk of radius R_f (Figure 6.8) there is no adaptation of the weights as the links are bidirectional, and therefore both the out- and the in-strength of a node are influenced by the addition of the new links, leading to a relative balance in flow. However, in the annulus $\mathcal{A}_{R_f}^{R_{\max}}(i)$, the links added are considered directional increasing the in-flow of the corresponding nodes, i.e., in Figure 6.8 the incoming flow of nodes $1, 6, 7$ has increased. Therefore, their out-strength has to be adapted so that they can satisfy the additional flow. Each one of the nodes in $\mathcal{A}_{R_f}^{R_{\max}}(i)$, where i is the newly added node, changes the weights of its out links as follows:

$$w'_{jh} = w_{jh} + \delta^{na} \frac{w_{jh}}{s_j^{out}}, \ j \in \mathcal{A}_{R_f}^{R_{\max}}(i), \ h \in (1, 2, .., N) \tag{6.15}$$

where δ^{na} is a constant parameter chosen as the corresponding weight variations in the WEC case, and depends on the application. Also, the weight adaptation, as in the WEC case, is considered local and concerns only the out-links of the corresponding nodes. As an example, in Figure 6.8, node 7 changes the weight w_{74} according to Eq. (6.15).

If considering each new link having a weight equal to the average of the probability distribution used for assigning weights to the new links and denoted as $E(P)$, then the change of the total weight, $w_a(t)$, of the network due to node addition is approximately equal to:

$$dw_a(t) = 2\pi M_a N(t) r \frac{R_f^2}{L^2} E(P) + \pi r M_a N(t) \frac{(R_f + at)^2 - R_f^2}{L^2} (E(P) + \delta^{na})$$

$$(6.16)$$

In Eq. (6.16), the first term corresponds to the bidirectional links in the range R_f. More specifically, the number of undirected links in the range R_f equals to $N(t)\pi \frac{R_f^2}{L^2}$ (Appendix A), and therefore this quantity is multiplied by 2 to obtain the number of directed links. The second term corresponds to the directional links added in each area $\mathcal{A}_{R_f}^{R_{\max}}(i)$ for each new node i. Also, $N(t)$ is the number of nodes in the network at step t.

Process of Node Deletion

At time slot t, node deletion takes place with probability v and M_d nodes, each one chosen with probability $\mathbb{P}_i^{nd}$ are deleted. The probability $\mathbb{P}_i^{nd}$ ($\sum_{i=1}^{N} \mathbb{P}_i^{nd} = 1$) is defined to be inversely proportional to some metric of participation of a node in the network. Specifically, a node with no strong connections with its neighbors is with high probability deleted. If a node is deleted then as expected, all its in- and out- connections are also deleted. Through the node deletion process, some unpopular nodes (according to the specific definition of unpopularity employed in each application) or some nodes with low participation in the network function can be replaced with others more popular or important and set more intelligently in the topology. In node deletion, there is no weight adaptation, since once a node is deleted, all of its connections are deleted bidirectionally as well. Considering that each deleted link has at $(t-1)$ step an average weight $\bar{w}(t-1)$ and the deleted node has average (in- or out-) degree $\bar{k}(t-1)$, then the total change of weight induced by node deletion is calculated as

$$dw_d(t) = -2v M_d \bar{k}(t-1)\bar{w}(t-1) \tag{6.17}$$

Weighted Node Churn (Addition and Deletion)

The Weighted Node Churn (WNC) mechanism consists of the node addition and deletion processes described above. WNC can be interpreted also as part of the natural evolution process of the network, where nodes leave and enter the network in its considered area. Its components, addition and deletion of nodes are two complementary processes as this is also implied by the natural network evolution. Essentially, the one cannot be considered without the other and vice-versa, since the first improves the network structure (average distance) and consequently the network performance, while the second ensures a balanced operation without excessive transmission costs and inter-ference. In the node churn case, at each time step t, node addition takes place with probability r, where M_a nodes are added. Node deletion takes place with probability v, where M_d nodes are deleted. Similarly with edge churn, $r + v \leq 1$. The average number of nodes at each time step t is theoretically equal to: $N(t) = N + M_a rt - M_d vt$ with N, the initial number of nodes in the network.

The average out-strength $\bar{s}^{out}$ (the same equation holds for $\bar{s}^{in}$), over all nodes, changes from step $(t-1)$ to the next step t as follows:

$$\bar{s}^{out}(t) = \bar{s}^{out}(t-1)\frac{N(t-1)}{N(t)} + 2\pi M_a r \frac{R_f^2}{L^2}E(P) + \qquad (6.18)$$
$$\pi r M_a \frac{(R_f + at)^2 - R_f^2}{L^2}(E(P) + \delta^{na}) - 2v M_d \frac{\bar{k}(t-1)\bar{w}(t-1)}{N(t)}$$

The first term of the right hand-side of Eq. (6.18) corresponds to the sum of the link weights over the network at step $(t-1)$, divided by $N(t)$, which is the number of nodes at the next step so as to obtain the averaged per node value at slot t. The other three terms reflect the mean change in the sum of link weights in the network expressed as an average over the $N(t)$ nodes. Specif-ically the sum of the last three terms of Eq. (6.18) is given by $\frac{dw_a(t) + dw_d(t)}{N(t)}$ ($dw_a(t), dw_d(t)$ are expressed in Eq. (6.16) and (6.17) correspondingly).

The total change in the number of links at each time step t may be ap-proximated as:

$$L_n(t) = \left(2M_a \pi r \frac{R_f^2}{L^2} + \pi r M_a \frac{(R_f + at)^2 - R_f^2}{L^2}\right)N(t) - 2v M_d \bar{k}(t-1) \quad (6.19)$$

The change in the number of links at time t is given by the sum $dw_a(t) + dw_d(t)$, when the weights of all links become equal to unity, i.e., in Eq. (6.16), (6.17), the values $E(P) = 1$, $\delta^{na} = 0$, $\bar{w}(t-1) = 1$ are assigned.

The approximate recursive equation of the average node degree is:

$$\bar{k}^{in}(t) = \bar{k}^{out}(t) = \bar{k}(t) = \frac{E + \sum_{k=1}^{t} L_n(k)}{N(t)} \qquad (6.20)$$

where E is the number of directed links in the initial RGG.

Assuming that W_s is the sum of the weights in the initial RGG topology and following the same logic as for obtaining Eq. (6.18), the change in total weight at time step t is:

$$dw(t) = 2\pi M_a N(t) r \frac{R_f^2}{L^2}E(P) + \pi r M_a N(t)\frac{(R_f + at)^2 - R_f^2}{L^2}(E(P) + \delta^{na})$$
$$-2v M_d \bar{k}(t-1)\bar{w}(t-1) \qquad (6.21)$$

Table 6.1: Combined mechanism.

Processes	Probability	Number of nodes or links participating
Edge addition	p	m_a links added
Edge deletion	q	m_d links deleted
Node addition	r	M_a nodes added
Node deletion	v	M_d nodes deleted

Therefore, the average weight in the network at time t is given by:

$$\bar{w}(t) = \frac{W_s + \sum_{k=1}^{t} dw(k)}{E + \sum_{k=1}^{t} L_n(k)} \qquad (6.22)$$

Thus, if the initial average connectivity ($\bar{k}(0)$), the initial average weight of the network ($\bar{w}(0)$), and the initial average strength of the network ($\bar{s}^{out}(0)$) are known, we can compute the average in- or out-strength by Eq.(6.18) with the aid of equations (6.20) and (6.22).

6.5.3   Combined Mechanism (WEC and WNC)

Finally, node churn and edge churn can be combined together and in this case the iTC-based topology modification mechanism consists of all the processes of the network evolution. It should be noted that the framework proposed to infuse social properties in wireless multi-hop networks takes into consideration their spatial character and does not impact their character as Random Geometric Graphs. In this case, the mechanisms and their probabilities are depicted in Table 6.1 (where $p + r + q + v \leq 1$):

The following section (Section 6.5.4) focuses on a methodology for obtaining optimal values for the parameters in Table 6.1 within a constrained optimization framework. Also, it is important to mention that the proposed mechanisms apply to a wide range of applications by properly specifying the probabilities $\mathbb{P}_i^a$, $\mathbb{P}_i^d$, $\mathbb{P}_i^{nd}$, $Q(k|i)$, $Q^d(k|i)$ and the values of the parameters δ^{1a}, δ^{2a}, δ^{La}, δ^{1d}, δ^{2d}, v_{ik} so as to fit in the framework of the specific application. For a realistic implementation of the proposed combined mechanism one should take into consideration the distributed nature of the wireless multi-hop network and the restricted energy resources by using optimal parameter values for each process (Table 6.1), so as to lead to the highest possible gain with the smallest energy consumption. As already mentioned, Section 6.5.4, provides a framework for optimizing the parameters of Table 6.1, given the permitted cost, which is related to the capabilities of the corresponding network and the desirable performance improvement. Finally, the schematic in Figure 6.9 summarizes the submechanisms of the holistic Topology Modification Mechanism, their relations, and their parameters that need to be specified for a possible practical application and development of the mechanism.

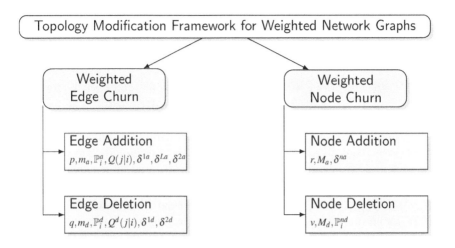

Figure 6.9: Topology modification mechanism binary tree diagram. The corresponding parameters of each process, subject to appropriate definition, are listed. The parameters' definition depends on the application that overlays the network, the required network performance, and the available network resources. The trade-off between performance, and resources consumption is analytically formulated and treated in the next section (Section 6.5.4).

6.5.4 Optimization Methodology

The topology modification mechanism described in Section 6.5 enables the enhancement of the wireless network topology with small-world properties, while it is able to adapt to different network functionalities and requirements through appropriate assignment of parameter values. However, the energy consumed by the modified nodes for each transmission increases polynomially with their transmission range. To alleviate the excessive increase in energy consumption, in [149] an optimization methodology was developed for designing the parameters defined in Section 6.5.3 and Table 6.1, in order to achieve the desired average path length, at the minimum additional cost regarding the energy consumption. Therefore, it is sought to optimize and control the parameter vector $\overrightarrow{P} = [p\ m_a\ q\ m_d\ r\ M_a\ v\ M_d]'$, in order to achieve an acceptable trade-off between the specified performance requirements, i.e., the needed average path length reduction, and the relevant cost paid, i.e., additional energy consumption due to radii increase. More specifically, it is possible to achieve a desired path length reduction by simply increasing the edge and node addition probabilities and at the same time turning off the deletion processes. However, this choice will lead to an excessive increase of energy consumption. On the other hand, by properly balancing the parameters of P, it will be shown that a desired path length reduction can be achieved while the increase in energy consumption is minimized. A detailed description of this optimization methodology can be found in [149].

Determination of the Objective Function

Towards this direction, a constrained optimization framework is developed with respect to $\overrightarrow{P}$, addressing the aforementioned trade-off. The employed objective function, which will be subject to minimization, should be related to the energy consumed by network nodes. To define the objective function, it is observed that given the transmission radius of each node and considering that each node transmits data (saturated network) and all messages have the same transmission time, a measure of the energy consumed by the network is $E(t, \overrightarrow{P}) = \sum_{i=1}^{N(t,\overrightarrow{P})} cR_i^\gamma(t, \overrightarrow{P})$, where γ is the path loss constant (depending on the communication medium ($\gamma \in [2, 4]$ as in reference [164] and others), c a constant [164], $N(t, \overrightarrow{P})$ is the number of nodes, and $R_i(t, \overrightarrow{P})$ the radius of node i, at time t with parameter vector $\overrightarrow{P}$. This is due to the fact that the transmission radius of node i to the path loss exponent γ is proportional to the node transmission power P_i^T, i.e., $P_i^T \sim R_i^\gamma(t, \overrightarrow{P})$, and thus an accurate indication of the consumed energy. However, obtaining the analytical form of the energy is difficult, since the analytical form of $R_i(t, \overrightarrow{P})\ \forall i = 1, ..., N(t, \overrightarrow{P})$ is required, which is rather complicated for a multi-hop network as well. However, at each time step t, the average radius $R_{avg}(t, \overrightarrow{P})$ satisfies the relation

$$\sum_{i=1}^{N(t,\overrightarrow{P})} R_i(t, \overrightarrow{P}) = N(t, \overrightarrow{P})R_{avg}(t, \overrightarrow{P}),$$

so that

$$\left(\sum_{i=1}^{N(t,\overrightarrow{P})} R_i(t, \overrightarrow{P}) \right)^\gamma = \left(N(t, \overrightarrow{P})R_{avg}(t, \overrightarrow{P}) \right)^\gamma,$$

yielding

$$\sum_{i=1}^{N(t,\overrightarrow{P})} R_i(t, \overrightarrow{P})^\gamma \leq \left(N(t, \overrightarrow{P})R_{avg}(t, \overrightarrow{P}) \right)^\gamma.$$

Therefore, given a maximum step t denoted as T, which equals with the time slots of the whole application of the iTC mechanism, the optimization of an upper bound of $E(T, \overrightarrow{P})$ can be achieved by minimizing the quantity $N(T, \overrightarrow{P})R_{avg}(T, \overrightarrow{P})$. Based on the iTC probabilistic framework developed in Section 6.5.3, an analytical expression of an upper bound of $N(T, \overrightarrow{P})R_{avg}(T, \overrightarrow{P})$ is possible to obtain.

Determination of the Equations for the Constraint Set

The constraint set should capture the desired small-world features of the final topology. Since analytical expressions of the average path length in a multi-hop network are hard to determine, even for simple random topologies, we are able to exploit the following observation regarding the iTC modification mechanism. A very tight relation exists between the average path length and the added shortcuts by the iTC approach. As verified in [86], by increasing the number of long-links added, the average path length decreases. Consequently, instead of conditioning on the average path length, one may condition on the number of shortcuts $L(T, \overrightarrow{P})$ added in the topology.

As also stated in Section 6.5.2, at each time step t, the average number of nodes in the network is $N(t, \overrightarrow{P}) = N + rM_a t - vM_d t$ and thus, the average transmission radius $R_{avg}(t, \overrightarrow{P})$ is:

$$R_{avg}(t, \overrightarrow{P}) = \frac{\sum_{i=1}^{N(t,\overrightarrow{P})} R_i(t, \overrightarrow{P})}{N(t, \overrightarrow{P})} = \frac{NR_f + dR(t, \overrightarrow{P})}{N(t, \overrightarrow{P})} \tag{6.23}$$

where $N(0, \overrightarrow{P}) = N$, NR_f is the sum of the initial node radii and $dR(t, \overrightarrow{P})$ is the total change of radii values across the network nodes up to step t. We have to compute the contribution of each node iTC sub-mechanism to $dR(t, \overrightarrow{P})$. Link deletion does not change the physical range of a node, thus has zero contribution. Node addition contributes at each time step k a quantity equal to $M_a r(R_f + ak)$, since at step k, with probability r, we add M_a nodes in the network, each one having a radius $R_f + ak$. Thus, up to step t, the contribution of node addition to $dR(t, \overrightarrow{P})$ is $M_a r \sum_{k=1}^{t} (R_f + ak)$. Concerning node deletion, its maximum contribution to $dR(t, \overrightarrow{P})$ up to step t is $-M_d v R_f t$, in the case that all nodes deleted have the initial radius R_f, defining an upper bound on $dR(t, \overrightarrow{P})$. Edge addition is more complicated as the network becomes heterogeneous with respect to node range, and the difference $R_{max}(t) - R_i(t, \overrightarrow{P})$, which expresses the increase of the transmission radius of selected node i at time t, depends on $R_i(t, \overrightarrow{P})$. In order to take into account the edge addition in the definition of the upper bound for $dR(t, \overrightarrow{P})$, we observe that the maximum value of $dR(t, \overrightarrow{P})$ occurs when all links are added to nodes with $R_i(t, \overrightarrow{P}) = R_f$. This value corresponds to an increment of the sum of radii that equals to $pm_a ak$ at step k (the explanation is the same as in the node addition case), and therefore this sums to $pm_a a \sum_{k=1}^{t} k$ up to step t.

Therefore, an upper bound for R_{avg} can be obtained of this probabilistic model:

$$R_{avg}^{up}(t, \overrightarrow{P}) = \frac{NR_f + M_a r R_f t + (M_a r + pm_a)a\frac{(t^2+t)}{2} - M_d v R_f t}{N + rM_a t - vM_d t} \tag{6.24}$$

Instead of $R_{avg}(t, \overrightarrow{P})$, we use in the objective function its upper bound equal to $R_{avg}^{up}(t, \overrightarrow{P})$. In the following, we calculate the average number of links added in the network. The number of links added up to step t is:

$$
\begin{aligned}
L(t, \overrightarrow{P}) = \ & pm_a t - qm_d t + \sum_{k=1}^{t} \left(2M_a \pi r \frac{R_f^2}{L^2} + \pi r M_a \frac{(R_f + ak)^2 - R_f^2}{L^2} \right) N(k) \\
& - 2M_d v \sum_{k=1}^{t} \pi \frac{R_f^2}{L^2} N(k-1)
\end{aligned} \tag{6.25}
$$

where, in the right-hand side, the first two terms correspond to EC and the last two summands to NC. Its derivation is similar to Eq. 6.19 of Section (6.5), including the WEC process. The coefficient 2 is used when the links added are bidirectional.

Optimization Formulation

The proposed approach minimizes an upper bound of the normalized energy consumption (by factor c), and consequently $N(t, \overrightarrow{P}) R_{avg}^{up}(t, \overrightarrow{P})$, under the following conditions:

1. The number of added links $L(t, \overrightarrow{P})$ has a fixed value ℓ_0, depending on the targeted value of mean path length, which ensures that the operation and the performance of the network is as desired.

2. The number of nodes $N(t, \overrightarrow{P})$ remains bounded by n_0, corresponding to a limit of the total active and non-active nodes (deletion may be considered temporal deactivation), or by the highest possible density of nodes in the topology.

3. The deletion process takes place less frequently than the addition, so as to avoid network disintegration.

4. Component-wise bounds $(\overrightarrow{LB}, \overrightarrow{UB})$ in $\overrightarrow{P}$ can be imposed for the probabilities of the sub-mechanisms and the number of nodes participating in each process.

The constraint set includes the necessary inequalities/equalities (i.e., conditions (1), (2), (3), (4) and a maximum t, denoted as T), in order to determine a realistic parameter set for the multi-hop network improving mechanism. Therefore, the following optimization problem can be obtained:

$$
\min_{\overrightarrow{P}} N(T, \overrightarrow{P}) R_{avg}^{up}(T, \overrightarrow{P}) \tag{6.26}
$$

$$
L(T, \overrightarrow{P}) = \ell_0, \ p + q + r + v = 1
$$

$$
N(T, \overrightarrow{P}) \leq n_0, \ p \geq q, \ r \geq v, \ m_a \geq m_d, \ M_a \geq M_d
$$

$$
m_a, m_d, M_a, M_d \in \mathbb{Z}, \ \overrightarrow{LB} \preceq \overrightarrow{P} \preceq \overrightarrow{UB}
$$

The proposed optimization setup has the special form of a separable optimization problem [25]. In the case of a separable optimization problem, as defined in [25], the objective function and the constraints can be written as sums of other sub-functions, each one being dependent only on a particular subset of optimization parameters, where the defined subsets of parameters are common for all the sub-functions. In this case, if two subsets of the parameters (i.e., $p, q, r, v, m_a, m_d, M_a, M_d$), $x_{EC} = \{p \; q \; m_a \; m_d\}$ and $x_{NC} = \{r \; v \; M_a \; M_d\}$ are defined, where the first one (x_{EC}) contains only edge churn parameters, while the second (x_{NC}) only node churn parameters, all the functions of the optimization problem can be expressed as sums of sub-functions, each one containing parameters only in x_{EC} or only in x_{NC}. These problems are often solved by using the dual[6] function, which eventually separates the problem into smaller minimization problems, each one depending only on a specific subset of the whole set of the parameters. In this case, the integer constraints are suitable for branch and bound optimization methods, so algorithmic/numerical evaluations can be effectively used. However, if duality is used, two minimization sub-problems would emerge (to build the dual function), one depending on x_{EC} and the other on x_{NC}. Therefore, this intuitively points to the fact that each of the mechanisms could be separately minimized, while satisfying all the constraints.

Simulation results shown analytically in [149] lead to the following basic interesting conclusions:

- **Observation 1:** There is an emerging trade-off between the Edge and Node Churn mechanisms. Edge Churn (link addition and deletion), if

---

[6]Let us suppose having the following optimization problem where f is a nonlinear function:

$$\min f(x)$$
$$\text{subject to } f_i(x) \leq 0 \; i \in 1, ..., m$$
$$h_i(x) = 0 \; i \in 1, ..., p \tag{6.27}$$

where $x \in D \subset \Re^n$. Let us denote with f^* the optimal value of this optimization problem. The Lagrangian function is defined as $L : \Re^n \times \Re^m \times \Re^p \to \Re$ and

$$L(x, \lambda, \nu) = f(x) + \sum_{i=1}^{m} \lambda_i f_i(x) + \sum_{j=1}^{p} \nu_j h_j(x),$$

where the variables $\lambda \geq 0 \in \Re^m$, $\nu > 0 \in \Re^p$ are denoted as dual variables or Langrange multipliers. The dual function is always concave irrespectively of the initial problem, and is defined as $h : \Re^m \times \Re^p \to \Re$:

$$h(\lambda, \nu) = \inf_{x \in \Re^n} L(x, \lambda, \nu) = \inf_{x \in \Re^n} \{f(x) + \sum_{i=1}^{m} \lambda_i f_i(x) + \sum_{j=1}^{p} \nu_j h_j(x)\}.$$

Finally, the dual problem is the following optimization problem:

$$h^* = \max_{\lambda, \nu} h(\lambda, \nu).$$

Regarding the relation between f^*, h^*, it always holds that $h^* \leq f^*$.

applied and optimized by itself, i.e., by turning off the Node Churn mechanism, leads to a higher optimal upper bound of the overall energy consumed, decreasing the clustering of the network faster, but it achieves a greater contraction of the hop distances in the network topology than in the case of Node Churn being applied and optimized by itself. As a result, Edge Churn needs more additional energy consumption and quickly decreases the degree of network clustering in favor of its effectiveness in reducing the average path length.

- **Observation 2:** The optimization method uses the lower possible bounds for the Edge Churn vector x_{EC} (when a feasible solution can be found) and adapts x_{NC} of the Node Churn mechanism, so as to achieve the desirable value of ℓ_0. This is intuitively expected if we take into consideration Observation 1 and the fact that the aim of the optimization is to reduce the additional energy consumption.

- **Observation 3:** The individual minimization of each EC, NC submechanism discussed previously, due to the problem separability, becomes evident from the attained equality of parameters of each process when possible (constraints do not permit more deletions than additions).

- **Observation 4:** Node Churn mechanism achieves a stable optimal upper bound of the overall energy consumed irrespective of the number of links added ℓ_0, or the maximum radius R_{MAX}. This observation is clearly evident in Figure 6.10. For Node Churn, the most suitable values of R_{MAX} and ℓ_0 will be chosen with respect to other criteria, such as the desirable mean path length, or the clustering coefficient.

- **Observation 5:** Although Edge Churn achieves higher minimum sum of radii (Figure 6.10), its participation in the whole topology modification mechanism, when desirable, should be increased through the constraints, and especially via the lower bound vector of $\vec{P}$, $\vec{LB}$, for the sake of performance improvement. Consequently, the number of links (shortcuts/contractions) added is important for the path length reduction, while the Edge Churn sub-mechanism should have a suitable percentage contribution in the modification mechanism, in order to obtain a desirable path length reduction with less added shortcuts. Node Churn, for its part, may need a lot of nodes added in the network to achieve individually a path length reduction of the order of the Edge Churn sub-mechanism. This trend can be observed in Figure 6.11.

Figures 6.10 and 6.11 verify some of the above observations, e.g., (1), (4), and (5). The observations (2) and (3) are illustrated in Figure 6.12. For more details, the interested reader should refer to [149]. In both Figures 6.10 and 6.11, each one of the two submechanisms Edge Churn and Node Churn is optimized separately so that relevant comparisons can be made. Figure 6.10

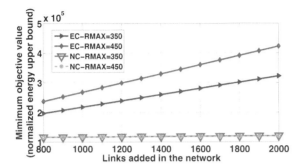

Figure 6.10: Comparison of the optimal sum of radii with respect to the number of links added in the network, for Edge Churn and Node Churn and two values of maximum radius $R_{\max} = 350$, $R_{\max} = 450$.

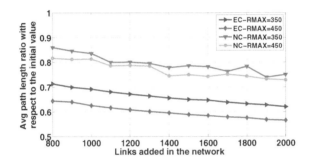

Figure 6.11: Comparison of the average path length with respect to the number of links added in the network, for Edge Churn and Node Churn and two values of maximum radius $R_{\max} = 350$, $R_{\max} = 450$.

presents the minimum objective value achieved by each of the mechanisms for the corresponding values of ℓ_0, $R_{\max}$. Similarly, Figure 6.11 presents the average path length improvement achieved by each one of the mechanisms Edge and Node Churn simulated with the optimal parameter vectors x_{EC} and x_{NC}, respectively, for the corresponding values of ℓ_0 and $R_{\max}$. In Figure 6.12, the optimal values of the holistic topology modification mechanism computed by the optimization problem defined by (6.26) are presented with $\overrightarrow{LB} = [0.4\ 10\ 0.01\ 1\ 0.01\ 1\ 0.01\ 1]'$ and for different values of ℓ_0. More specifically, Figure 6.12(a) depicts the optimal values of the parameters of the probability distribution, p, q, r, v (Appendix A), and Figure 6.12(b) depicts the optimal values of the parameters of the number of nodes/edges being added/deleted, m_a, m_d, M_a, M_d.

As an example, supposing that we need a high average path length reduction with the restriction that the addition of a high number of nodes/hubs is

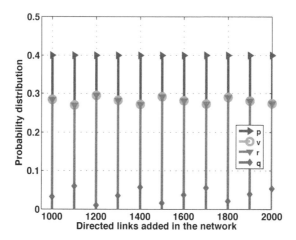

(a) Probability distribution, $R_{MAX} = 350m$.

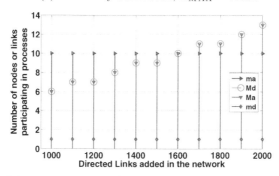

(b) Numbers of nodes/links participating in the processes, $R_{MAX} = 350m$.

Figure 6.12: Optimal parameter values.

not possible. One way to achieve this is to increase the participation of the edge addition process by determining appropriately the vectors $\overrightarrow{LB}$ of Problem (6.26), so that we can achieve the desired low path length by increasing the transmission range of already existing nodes and avoiding overcrowding the network with new nodes. Then the solution of the Problem (6.26) will assign the lower bound parameters to the edge addition process and adapt the rest of the processes in order to satisfy the constraints. In this case the optimal (minimum) increase in energy consumption will be higher, compared to the case of assigning a zero valued lower bound on the edge addition process. This conclusion means that if $LB = [0\ 0\ 0\ 0\ 0\ 0\ 0\ 0]'$, it is optimal in terms of energy consumption to perform topology modification by using only the Node Churn process.

6.6 Special Cases

6.6.1 Example 1: Elimination to Binary Graphs (SETM)

At this section, the general topology modification mechanism is applied in the case of directed, unweighted graphs, i.e., the weights of the existing links are all equal to 1. In this case, the topology modification mechanism is denoted as "Socially Evolutionary Topology Modification Mechanism (SETM)" due to both the infusion of small-world properties in the wireless network topology and the use of preferential attachment for the definition of the probabilities $\mathbb{P}_i^a$, $\mathbb{P}_i^d$, $\mathbb{P}_i^{nd}$, $Q(k|i)$, $Q^d(k|i)$, i.e., the topology modification mechanism has a socially-inspired design concept. Its detailed presentation can be found in [148]. Since the weights are equal to unity, the node strengths coincide with the node degrees, i.e., $s_i^{out} = k_i^{out}$, $\forall\ i$, and $s_i^{in} = k_i^{in}$, $\forall\ i$. Also, as mentioned above, $w_{ij} = 1$ if the link (i,j) exists, and similarly for a newly added link (i,k) we assign weight $v_{ik} = 1$. There is no reason for weight adaptation, therefore, $\delta^{1a} = 0$, $\delta^{2a} = 0$, $\delta^{La} = 0$, $\delta^{1d} = 0$, $\delta^{2d} = 0$. For SETM, Eq. (6.13), (6.7) and (6.18) become mathematically tractable and analytic solutions can be obtained.

Regarding Edge Addition, the first endpoint of the link is selected with preferential attachment, i.e., proportionally to the node out-degree.

$$\mathbb{P}_i^a(k_i^{out}) = \frac{k_i^{out} + c}{\sum_{\text{all nodes } j} (k_j^{out} + c)} \qquad (6.28)$$

As a result, the nodes with high degree are more likely to be selected, while a node with zero degree can be still chosen, due to the constant factor (here $c = 1$). The other endpoint k is a randomly chosen neighbor, among the new ones in the annulus $A_{R_{\min}}^{R_{\max}}(i)$. Such probability is equal to the probability that the node is in the ring area $A_{R_{\min}}^{R_{\max}}(i)$ (area $|B|$) of node i, which equals (Appendix A):

$$Q(k|i) = \frac{|B|}{L^2} = \frac{\pi(R_{\max}^2 - R_{\min}^2)}{L^2}.$$

Regarding Edge Deletion, the first endpoint i of the deleted link is selected with probability

$$\mathbb{P}_i^d(k_i^{out}) = \frac{N - 1 - k_i^{out} + c}{\sum_{\text{all nodes } j} (N - 1 - k_j^{out} + c)} \qquad (6.29)$$

and the second endpoint k is a random one-hop neighbor of i, i.e., selected with probability $Q^d(k|i) = \frac{1}{k_i^{out}}$. The first endpoint of the deleted edge is chosen with inverse preferential attachment probability, i.e., with probability inversely proportional to the popularity of a node. Inverse preferential attachment is obtained by subtracting the degree of each node k_i^{out} from the maximum possible degree $N-1$ (if the number of nodes in the network is N), which represents an effective node degree value for inverse preferential attachment and then replacing such quantity in the preferential attachment rule. The constant $c=1$ has a role similar to that in probability $\mathbb{P}_i^a(k_i^{out})$, given in (6.28).

The total rate of change of the node out-degree is the sum of the right-hand sides of Eqs. (6.13), (6.7) as follows:

$$\frac{ds_i^{out}}{dt} = \frac{dk_i^{out}}{dt} = pm_a\mathbb{P}_i^a(k_i^{out}) - rm_d\mathbb{P}_i^d(k_i^{out}) \qquad (6.30)$$

The form of the solution may be obtained as:

$$k_i^{out}(t) = e^{(f(t)+g(t))}h(t) + e^{g(t)}w(t)d \qquad (6.31)$$

where the functions f, g have the form: $f(t) = A\arctan(Bt + \ell)$, $g(t) = A\arctan(-Bt - \ell)$, where A, B and ℓ are defined constants and $h(t), w(t)$ are defined polynomial functions. The value of d is obtained by the initial condition $k_i(0) = \overline{C}$, where $\overline{C}$ denotes the initial average node degree.

Regarding node deletion, we consider that the selection of a node for being deleted is equal to $\mathbb{P}_i^{nd} = \mathbb{P}_i^d(k_i)$, by replacing N with $N(t)$ since when introducing node churn, the number of nodes changes with t.

In the case of binary graphs, Eq. (6.18) becomes as follows:

$$\bar{k}^{out}(t) \cong \bar{k}^{out}(t-1)\frac{N(t-1)}{N(t)} + 2\pi M_a r\frac{R_f^2}{L^2}$$

$$+\pi r M_a\frac{(R_f + at)^2 - R_f^2}{L^2} - 2vM_d\frac{\bar{k}(t-1)}{N(t)} \qquad (6.32)$$

since one can make the following replacements: $E(P) = 1$, $\delta_{na} = 0$, $\bar{w}(t-1) = 1$, $\frac{\bar{k}(t-1)}{N(t)} \cong \pi\frac{R_f^2 N(t)}{L^2 N(t)}$ where the last approximate equality is due to the choice of nodes for deletion with inverse preferential attachment, $\mathbb{P}_i^d(k_i^{out})$, which assigns high probability to nodes with the lowest transmission radius R_f to be selected for deletion.

Based on Eq. (6.32), the total rate of change of the average connectivity k_i^{out} of each node i is given by the differential equation:

$$\frac{dk_i^{out}}{dt} = 2\pi M_a r\frac{R_f^2}{L^2} + \pi r M_a\frac{(R_f + at)^2 - R_f^2}{L^2} - vM_d\pi\frac{R_f^2}{L^2} \qquad (6.33)$$

The solution of the above is: $k_i^{out}(t) = \frac{2R_f^2 M_a r\pi t}{L^2} - \frac{R_f^2\pi vM_d t}{L^2} + \frac{aR_f M_a r\pi t^2}{L^2} + \frac{a^2 M_a r\pi t^3}{3L^2} + \overline{C}$ where $\overline{C}$ is the average node degree of the initial network.

We should mention that the above specification of the holistic topology modification mechanism for unweighted graphs is not unique, but rather it can take other forms as well, in order to serve different applications. Another possible consideration would probably have expressed the probability $\mathbb{P}_i^a$ proportional to a node's degree at the social layer, i.e., with the neighbors of a node at the application/social layer and not at the physical layer. Therefore, a node having many friends to communicate with (this communication may require multiple hops in the physical layer) would have a higher transmission range at the physical layer. This adaptation expresses and utilizes the socio-physical layer interaction, by designing the physical layer according to the social layer and by taking into consideration the restrictions of the physical layer.

In the following, the infusion of small-world properties in the topology of wireless multi-hop networks, modeled by a binary (unweighted) graph, in the case of Edge Churn, Node Churn and SETM, the latter being the combined Edge and Node Churn, is demonstrated. The network topology consists of $N = 750$ nodes, $L = 2000m$ and $R_{MIN} = R_f = 150m$. At each step, the current value at time t of $R_{\max}(t)$ increases progressively, reaching $R_{MAX} = 450m$ eventually. The value by which the transmission radius $R_{\max}(t)$ increases at each step, a, is appropriately chosen, so that at least one node exists in the extended annulus of the selected node i, $\mathcal{A}_{R_{\min}}^{R_{\max}}(i)$, based on the network density.

Figure 6.13(a) exhibits the behavior of the average path length and clustering coefficient for the execution of the proposed Edge Churn iTC mechanism. The average path length decreases as expected, due to the addition of shortcuts in the local neighborhood of selected nodes. Even though one might expect that addition of new links leads to increasing clustering coefficient scaling, on the contrary, the latter decreases. The clustering coefficient in the induced digraph is evaluated in the same manner as in an undirected graph (i.e., by considering the edges undirected). Edge additions happen in random places and not in a specific area leading to the reduction of the clustering coefficient in Eq. (4.5) of Chapter 4, since each added link is randomly placed, rather than targeted around specific nodes or areas. Figure 6.13(b) presents the out-degree distribution in the initial and induced networks. The node degree distribution (Appendix A) of the initial network shifts towards higher values (which is reflected in the average path length reduction), while it maintains the random geometric nature of the network. Both distributions are of similar shape. However, the distribution of the induced graph contains more probability mass towards higher node degrees and it stochastically dominates[7] (first order domination) the initial node degree distribution [93].

---

[7] Let us consider two discrete probability distributions $\hat{P}$ and P. The main idea behind the definition of the first-order stochastic dominance is that P is obtained by shifting mass from $\hat{P}$ to place it on higher values. More specifically, the following conditions are equivalent:

$$\sum f(d)P(d) \geq \sum f(d)\hat{P}(d) \text{ for all nondecreasing functions } f$$

$$\sum_{d=0}^{x} P(d) \leq \sum_{d=0}^{x} \hat{P}(d) \text{ for all } x$$

$$\sum_{d=x}^{\infty} P(d) \geq \sum_{d=x}^{\infty} \hat{P}(d) \text{ for all } x$$

If these conditions hold, we say that the distribution P first order stochastically dominates $\hat{P}$ [93].

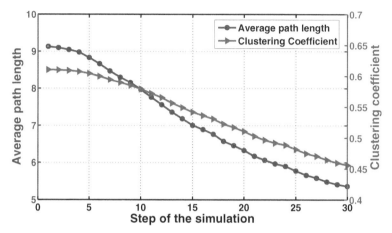

(a) Scaling of mean path length and clustering coefficient.

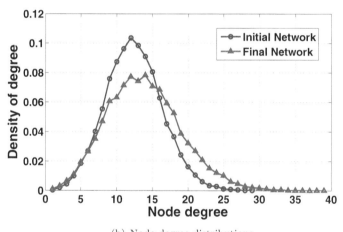

(b) Node degree distributions.

Figure 6.13: Edge Churn: Scaling of metrics. The parameters of the Edge Churn process are $p = 0.8$, $m_a = 75$ for addition, $q = 0.2$, $m_d = 45$ for deletion, and $a = 10m$.

In the following the demonstration of the node churn process is provided by using the same initial network topology, where at each time t the parameters of node addition and node deletion are chosen as $r = 0.7$, $M_a = 5$, and $v = 0.3$, $M_d = 1$, correspondingly.

Figure 6.14(a) exhibits the decrease in the average path length and the behavior of the clustering coefficient at each time step. It can be observed that the average path length is reduced as expected, because of the addition of nodes with increased radius. The clustering coefficient (Eq. (4.5) of Chapter

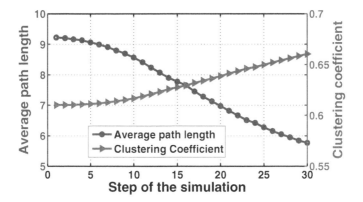

(a) Scaling of mean path length and clustering coefficient.

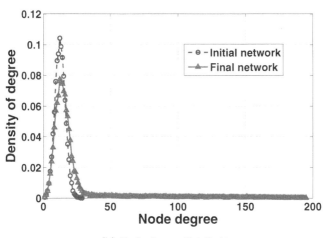

(b) Node degree distributions.

Figure 6.14: Node Churn: Scaling of metrics.

4) increases, because now new edges are added in localized areas around the new nodes, augmenting the connectivity percentage of these areas. The scaling behavior of the clustering coefficient yields stronger connectivity between neighboring nodes, and retains the initial character of the RGG. Figure 6.14(b) demonstrates the node degree distributions of the initial and induced networks. Similar observations as with Edge Churn apply. The induced network maintains the features of the initial. However, in the induced network, significant mass of the distribution shifts towards the higher degree values leading to a larger tail, characteristic of scale-free networks (Chapter 5). The greatest percentage of it remains close to the previous concentration area, and a small percentage shifts to greater values of node degree. In comparison to

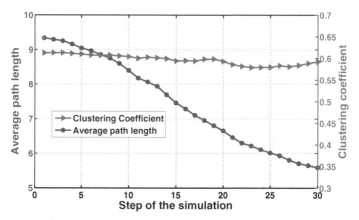

(a) Scaling of mean path length and clustering coefficient.

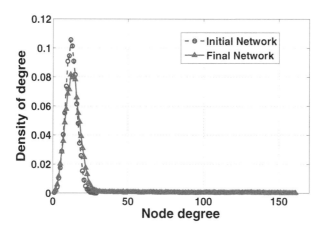

(b) Node degree distributions.

Figure 6.15: SETM: Scaling of metrics.

Edge Churn modification mechanism, the maximum degree is higher, but the number of nodes having connectivity higher than 50 is limited.

Finally, in order to check how Edge and Node Churn wrap up together, the operation for SETM mechanism for a similar network topology is demonstrated. The following parameters were used for the simulation of the SETM mechanism: $p = 0.4$, $m_a = 75$, $q = 0.15$, $m_d = 45$, $r = 0.4$, $M_a = 5$, $v = 0.15$, $M_d = 3$. Each combination of parameters yields a different result with respect to the average path length and clustering coefficient.

Figure 6.15(a) exhibits the decrease in the average path length, and the behavior of the clustering coefficient at each step. Consequently, by implementing both Node and Edge Churn with the aforementioned choice of parameters, the average path length of the network can be significantly reduced, without altering the original nature of the network (by retaining as almost constant the initial value of the clustering coefficient), which allows the network to operate properly for its designated mission. Figure 6.15(b) demonstrates the node degree distributions of the initial and induced topologies. Similar observations as before apply (i.e., for the case of Edge and Node Churn), i.e., the induced degree distribution stochastically dominates the initial in the first order sense and it is characterized by a long tail, which confirms the topological heterogeneity of the network, as a consequence of the addition of nodes and links.

At this point, betweenness centrality (Chapter 4) is used as a measure of the amount of flow that a node controls or as a measure of the congestion of each node. This is a common use of betweenness centrality in communication networks, since by definition, betweeness centrality coincides with the percentage of paths passing through a node in the network and thus, it can be easily translated to the percentage of packets passing through a node over the total number of packets exchanged in the network. Similarly to the definition of node betweenness centrality, edge betweenness centrality can also be defined (Chapter 4), for every edge instead of every node. At this point the centrality (based on betweenness) of the different categories of nodes in the network such as the existing nodes, or nodes that have added shortcuts due to Edge Churn, or nodes/hubs that have been added during the node addition process, is studied. Regarding Node Churn, Figure 6.16 shows that the newly added nodes have the highest betweenness centrality (Figure 6.16(a)) in the final network, while their betweenness centrality decreases as the number of nodes M_a, or the probability r increase (Figure 6.16(b)). Consequently, when applying Node Churn iTC, a sufficient number of nodes should be added, so that the traffic load is balanced and bottlenecks are avoided when the network traffic is heavy. Similarly, for the Edge Churn induced network, the betweenness centrality per edge (Figure 6.16(c)) and per node (Figure 6.16(d)) are depicted and appear to be more homogeneous among the edges and nodes correspondingly.

The proposed mechanisms lead to infusion of social properties in the wireless network topology for properly chosen parameters. Both churn mechanisms reduce the average path length, while node churn helps at keeping the clustering coefficient stable compared with the initially highly clustered ad hoc topology, as in small-world networks. In [148], SETM is studied under a realistic implementation in a wireless multi-hop network, where metrics such as throughput and delay have been counted in real data traffic between node pairs. An important improvement in the network performance is observed for both delay (decrease up to 30%) and throughput (increase up to 40%), without sacrificing much energy, as the energy consumption is increased by at most 4%. This topology improvement can be combined with efficient scheduling and routing algorithms for further improving the performance of wireless multi-hop networks.

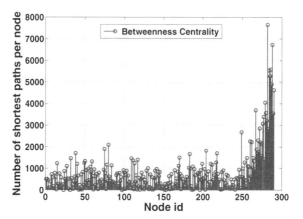

(a) Node Churn network.
Betweenness centrality per node.
Initial nodes: $id \leq 250$,
New nodes: $id > 250$, $M_a = 5$, $r = 0.7$, $v = 0$.

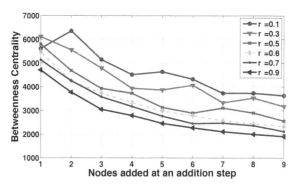

(b) Node Churn network. Average node betweenness centrality.

Figure 6.16: Betweenness centralities ($L = 800m$, $N = 250$, $R_f = R_{MIN} = 120m$, $R_{MAX} = 220m$, $a = 10m$).

6.6.2 Example 2: Trust Management in Wireless Multi-Hop Networks

This example shows the potential of designing the holistic topology modification mechanism via its parameters in order to take into consideration the weights of the network graph. We focus on the case that the weights represent trust values and the topology modification mechanism's parameters are selected as being functions of the trust weights. Commonly used trust models are mainly based on a weighted and directed graph model, in order to denote

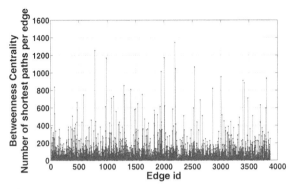

(c) Edge Churn network.
Betweenness centrality per
edge ($m_a = 25$, $p = 0.7$, $q = 0$).

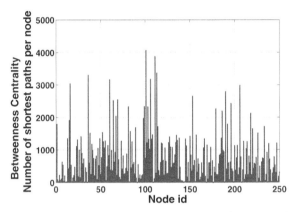

(d) Edge Churn network. Betweenness centrality per node
($m_a = 25$, $p = 0.7$, $q = 0$).

Figure 6.16: (Continued)

trust relations between node pairs (personal beliefs, statistical recommenda-
tions, etc.) [95, 153]. Two interacting nodes i, j, may assign local trust values
w_{ij}, w_{ji} expressing their biased trust perception. These two values may differ,
as the opinion of i for the trustworthiness of j may be different from the opinion of j for i. The obtained weighted directed graph describing such relations
is called trust graph. In the static wireless multi-hop networks considered in
our work, the trust graph coincides with the physical topology due to the
locality of wireless communications and the weight matrix W expresses the
trust values for all one-hop node pairs. In this work, a trust value equal to 0
means that node i has no direct communication with node j.

Framework Specification and Operation

Such a type of network, i.e., where the weight matrix W represents the trust graph, can be used in cases where trusted communication is more important than shortest path communication. However, the most trusted path between a node pair might be much longer in hops compared to the corresponding shortest path. The application of the topology modification mechanism, while incorporating the trust weight values through the selection of its parameters, can lead to contracting the most trusted paths allowing for faster communication with better quality. Taking this into account, in this section, the proposed topology modification framework is adapted by determining the probabilities, $Q(k|i)$, $Q^d(k|i)$, $\mathbb{P}^a_i$, $\mathbb{P}^d_i$, $\mathbb{P}^{nd}_i$, and aiming at jointly reducing the mean hop count of the most trusted paths, in addition to increasing the corresponding path trust value. We use a simple definition for the total trust value of a path, as the multiplication of link weights of a path. Thus, the most trusted path separating two nodes, is the one possessing the highest trust value. This value is denoted as "distance" in the trust graph. The weights are normalized in the interval $[0,1]$ for simplicity. Based on the above definition of the trust value of a path and the most trusted path, the semiring $(R, \oplus, \otimes, 0, 1) \equiv ([0,1], \max, \times, 0, 1)$ (Appendix B) is used for computing the corresponding quantities.

In the deletion process, a node i should choose with high probability a distrusted out-neighbor k to remove from its neighbor list, according to probability $Q^d(k|i)$, which expresses inverse preferential attachment with respect to the weight w_{ik}:

$$Q^d(k|i) = \frac{w_{\max} - w_{ik}}{\sum_{j \in \mathcal{N}^{out}(i)} (w_{\max} - w_{ij})} \tag{6.34}$$

A node k with low value of w_{ik} is more probable to be selected to lose its in-link from node i. The value of $w_{\max}$ is equal to 1 in this case.

At this point, we provide two probability based selection rules. According to the first one, denoted as in-strength preferential attachment, selection is performed proportionally to the total weight entering node i and is expressed by:

$$\Pi^{in}_i = \frac{s^{in}_i}{\sum_{j=1}^N s^{in}_j} \tag{6.35}$$

while the second one is inversely proportional to the in-strength, namely:

$$\Pi^{invs}_i = \frac{\frac{1}{s^{in}_i + \theta}}{\sum_{j=1}^N \left(\frac{1}{s^{in}_j + \theta}\right)} \tag{6.36}$$

Π^{invs}_i is obtained from Eq. (6.35), by replacing s^{in}_i with the inverse in-strength equal to $\frac{1}{s^{in}_i}$. As s^{in}_i decreases, $\frac{1}{s^{in}_i}$ increases, leading to higher selection probability for low-strength nodes. The positive parameter θ (taking very small values) ensures the rule works even for disconnected nodes with zero strength. Similarly, it can be defined the out-strength (inverse) preferential attachment by replacing, in Eq. (6.35) and (6.36), Π_i^{in}, Π_i^{invs}, and s_i^{in} with Π_i^{out}, Π_i^{outvs}, and s_i^{out}, respectively. Strength preferential attachment generalizes the degree-driven preferential attachment for weighted networks (Section 6.6.1).

In addition, $\mathbb{P}_i^a$ is identified with Π_i^{in}, based on the observation that a node with high in-strength is trusted by its one-hop neighbors and thus, more preferable to participate in the creation of shortcuts in the trust graph. Similarly for edge deletion, $\mathbb{P}_i^d$ is chosen as Π_i^{invs}, so as to avoid distrusted nodes included in paths. On the contrary, the rule $\mathbb{P}_i^{nd}$ will delete nodes having low trusted connections, i.e., with probability inversely proportional to their out-strength, thus $\mathbb{P}_i^{nd} = \Pi_i^{outvs}$. It can be observed that when $\mathbb{P}_i^a = \Pi_i^{in}$, the creation of chains of long-range connections may take place. This is possible, if for instance the link $i \to j$ is formed and thus, s_j^{in} increases, leading to a higher probability that j will be chosen at a next step.

In the sequel, two variations of the probability $Q(k|i)$ that the shortcut $i \to k$ is formed are proposed. In the first one, referred to as a local algorithm, nodes require local information only in order to perform edge addition, while in the second, denoted by global algorithm, nodes require a network wide view. We define $Q(k|i)$ in two possible ways:

- Local algorithm: In this case, node i requests information from its local neighbors (1 or 2 hops away), in order to learn the trust values of its potential neighbors in $\mathcal{A}_{R_{\min}}^{R_{\max}}(i)$, so that eventually the long link added is limited to 2 or 3 hops. Let h_2 be the nodes in $\mathcal{A}_{R_{\min}}^{R_{\max}}(i)$ that are 2 hops away from i and h_3 the nodes 3 hops away from i.

  - 2 hops: If node k is 2 hops away from i, its probability of being selected to form the link $i \to k$ is:

  $$Q(k|i) = \frac{\sum_h w_{ih} w_{hk}}{\sum_{h,(m \in h_2)} w_{ih} w_{hm} + \sum_{h,g,(m \in h_3)} w_{ih} w_{hg} w_{gm}}$$
  $$= \frac{W_{(ik)}^2}{\sum_{m \in h_2} W_{(im)}^2 + \sum_{m \in h_3} W_{(im)}^3} \tag{6.37}$$

  The weight assigned to the long-link is:

  $$v_{ik} = \max_h (w_{ih} w_{hk}) \tag{6.38}$$

  - 3 hops: If k is 3 hops away from i, its probability of being selected is:

$$Q(k|i) = \frac{\sum_{h,g} w_{ih} w_{hg} w_{gk}}{\sum_{h,(m\in h_2)} w_{ih} w_{hm} + \sum_{h,g,(m\in h_3)} w_{ih} w_{hg} w_{gm}} \tag{6.39}$$

$$= \frac{W^3_{(ik)}}{\sum_{m\in h_2} W^2_{(im)} + \sum_{m\in h_3} W^3_{(im)}} \tag{6.40}$$

$$v_{ik} = \max_{h,g}(w_{ih} w_{hg} w_{gk}) \tag{6.41}$$

- Global algorithm: In this case there is no limitation on the hops covered by the added long link (depending however on the power capabilities of the nodes), while node i is assumed to obtain information from the whole network aiming to select with high probability the most trusted node in $\mathcal{A}^{R_{max}}_{R_{min}}(i)$. In this case, the node i should compute the path semirings (Appendix B) for all nodes in $\mathcal{A}^{R_{max}}_{R_{min}}(i)$ and the probability that $k \in \mathcal{A}^{R_{max}}_{R_{min}}(i)$ is selected by node i is given as follows:

$$Q(k|i) = \frac{D^t_{i\to k}}{\sum_{j\in\mathcal{A}^{R_{max}}_{R_{min}}(i)} D^t_{i\to j}} \tag{6.42}$$

$$v_{ik} = D^t_{i\to k} \tag{6.43}$$

where $D^t_{i\to k}$ is the distance between i and k in the trust graph as noted above. In both algorithms node i chooses with high probability a node k for which the path $i \to k$ has a high trust value.

Evaluation and Discussion

This section clarifies through simulation results the benefits in the network topology and performance by applying a topology modification mechanism tied with the network trust values. The simulation results concern the average path length (geodesic) in hops, the trust value of the most trusted paths (distance), the number of hops of the most trusted paths and the clustering coefficient. An initial weighted directed RGG of $N = 250$ nodes, $R_f = 100m$, square deployment area of side $L = 800m$, $R_{MAX} = 250m$, and $a = 10m$ is considered. Initially, positive weights are assigned randomly on the directed links. Weight adaptation does not have a special meaning in trust graphs, as the direct opinion of a node for its neighbors does not change with edge variations. Therefore, it is considered $\delta^{1a} = \delta^{2a} = \delta^{1d} = \delta^{2d} = \delta^{na} = 0$.

If only edge churn is performed, the parameters used are: $p = 0.7$, $q = 0.3$, $m_a = 25$, $m_d = 8$. In the case of node churn, the parameters used are $r = 0.7$, $v = 0.3$, $M_a = 3$, $M_d = 1$. For the combined mechanism, the parameters used for each process are $p = 0.35$, $m_a = 25$, $q = 0.15$, $m_d = 8$, $r = 0.35$, $M_a = 3$, $v = 0.15$, $M_d = 1$.

In Node Churn, shown in Figure 6.17(a), the newly added nodes are considered as trusted and the weights of their in-links are assigned according to

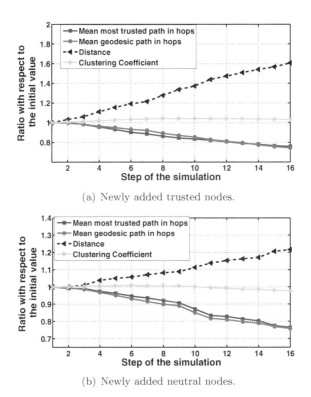

(a) Newly added trusted nodes.

(b) Newly added neutral nodes.

Figure 6.17: Node Churn mechanism performance for trusted and neutral newly added nodes.

the beta $\beta(5,1)$ distribution, which gives with high probability values close to 1, and of their out-links, according to $\beta(2,2)$ which gives with high probability values close to 0.5. In Figure 6.17(b) the weights are randomly chosen from $\beta(2,2)$ to both the in- and out-connections of the newly added nodes, so as to consider their trustworthiness as neutral. It can be observed in Figure 6.17 that the Node Churn mechanism achieves both the reduction of the mean geodesic path (in hops) and of the mean most trusted path (in hops) at approximately the same percentage. In addition, it maintains a stable clustering coefficient. The links of the new nodes with higher ranges serve as shortcuts, while nodes with distrusted connections are deleted, leading to higher values of most trusted paths (distances). When the newly added nodes are highly trusted, the hop number of the most trusted paths reduces with higher rate and the average distance increases more (Figure 6.17(a)).

Two cases of the parameter δ^{La} are examined, i.e., $\delta^{La} = 0$, $\delta^{La} = 0.1$. Initially, Figure 6.18 presents the results when weight adaptation is not considered ($\delta^{La} = 0$). In Edge Churn, the most trusted paths and the geodesic paths decrease for both local and global algorithms. However, the global al-

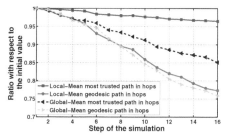

(a) Edge Churn (geodesic and most trusted paths).

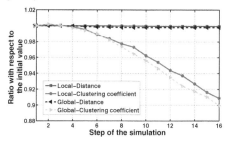

(b) Edge Churn (distance and clustering coefficient).

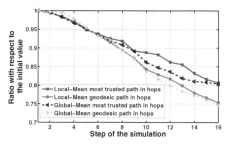

(c) Combined Mechanism (geodesic and most trusted paths).

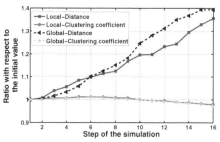

(d) Combined Mechanism (distance and clustering coefficient).

Figure 6.18: Performance of the proposed framework—comparison of local and global algorithms for the Edge Churn and combined mechanism in case $\delta^{La} = 0$.

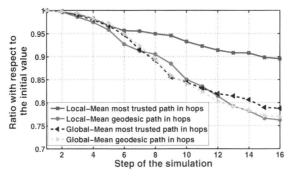

(a) Edge Churn (geodesic and most trusted paths).

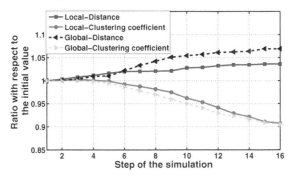

(b) Edge Churn (distance and clustering coefficient).

Figure 6.19: Performance of the proposed framework—comparison of local and global algorithms for the Edge Churn (with weight adaptation $\delta^{La} = 0.1$).

gorithm leads to higher reduction as far as the number of hops of the most trusted paths is concerned, due to the more complete knowledge of the network it has available. In the case of the global algorithm the new link $i \rightarrow k$ attains the value v_{ik} of the most trusted path between i and k, while in the local algorithm $v_{ik} = \max_h(w_{ih}w_{hk})$ or $v_{ik} = \max_{h,g}(w_{ih}w_{hg}w_{gk})$, which is not surely the highest value of trust between i and k. For the global algorithm, the semiring path algorithm will choose either the multi-hop most trusted path or the new shortcut in computing the path between two nodes. Both of them have the same trust value, and thus the path algorithm does not ensure the selection of the one hop shortcut, so that eventually it cannot achieve the largest reduction in the mean most trusted path in terms of hops. For the local algorithm, the value v_{ik} may not be equal to $D^t_{i \rightarrow k}$ and thus, not selected by the semiring based algorithm. The clustering coefficient is reduced due to the addition of long links, which in turn reduce the proportion of triangles to triplets (if one considers the corresponding definition of clustering coefficient). Also, the mean distance remains stable, as we do not perform weight adaptation.

The combined mechanism exploits the characteristics of both the Edge Churn and Node Churn mechanisms. The new nodes are considered as trusted. The clustering coefficient is approximately stable, the distance is increased, and the influence of the locality in the local algorithm is reduced due to the Node Churn mechanism. Thus, the geodesic paths are reduced, and the number of hops of the most trusted paths is also reduced with a percentage close to this of the geodesic paths. As can be observed, for this choice of parameters, the results of the local algorithm are close to those of the global algorithm leading to the choice of the local algorithm for less overhead.

Figure 6.19 shows that a low value of $\delta^{La} = 0.1$ makes the improvements more evident, especially for Edge Churn. Specifically, for the global algorithm it distinguishes the new link from the corresponding multi-hop path with the same trust value, while for the local algorithm the increase in v_{ik} will probably render the weight of the long link higher than $D^t_{i \to k}$, helping the path algorithm to select the new shortcut. The distance is increased as expected, due to the increment of the weights of the long-range links and both the local and the global algorithms lead to higher reductions in the number of hops of the most trusted paths. The mean geodesic path does not change much compared to the case of $\delta^{La} = 0$, as expected.

6.7   Conclusions

In this chapter we discussed and developed the idea of enhancing wireless multi-hop networks with social networks' properties aiming to improve their performance regarding delay, throughput, and average path length. We presented evolutionary topology modification mechanisms that incorporate appropriate shortcuts in the topology of wireless multi-hop networks targeting at creating small-world like structure in the topology of spatial wireless multi-hop networks. Firstly, we described diverse topology modification mechanisms from the related bibliography by dividing them in two categories depending on the type of shortcuts they use, wireless or wired. For most of them we provided indices for the achieved performance improvement, either theoretically or through simulation/numerical results. Secondly, we focused on a general topology modification mechanism, consisting of two basic processes, namely Edge Churn and Node Churn. This mechanism can infuse small-world properties in a non-dense network through inverse Topology Control by a proper selection of its parameter vector. In addition, through an optimization framework, its parameter vector can be adapted so as to balance the trade-off between cost and performance improvement. Two implementations of the holistic topology modification mechanism, one in binary and one in weighted networks are analytically presented. The holistic topology modification mechanism presented is based on two mechanisms characterizing the natural network evolution (edge and node churn), i.e., they can take place in the network topology via the physical process of network development, not only through biased control. Finally, its application is not restrictive in wireless multi-hop networks, but by making appropriate modifications (i.e., relax the spatial constraints), it may be applied in other networks, such as wired networks or relational social graphs, e.g., as a network of hospitals or airports, so as to improve the medical care or the transportation efficiencies, correspondingly.

Chapter 7

Conclusion

In the previous chapters of this book, we progressively developed and presented a framework for classifying, studying, analyzing, and designing evolutionary algorithms for complex networks, with special emphasis on wireless decentralized networks. Adopting the viewpoint of Network Science and utilizing elements from social/complex network analysis, novel approaches in the analysis and design of wireless complex networks were presented from a broader perspective enabling their analysis and further extension. The main purpose was to provide a working viewpoint of the corresponding methodology and to demonstrate the potentials of the proposed approaches, while also explaining the benefits of further adopting this inherently inter-disciplinary and multi-disciplinary framework for the analysis and design of these and other types of networks.

In this last chapter, we will first summarize the most important highlights of the previous ones, in order to make clearer the benefits of employing social/complex network analysis in modern communications networks research and then spend some time to indicate some interesting open issues constituting smaller or higher streak directions for future research endeavors in the corresponding research area. Finally, we conclude this book by providing a holistic current and future picture of the framework of evolutionary network analysis and design, which will shed additional light in the understanding and adoption of the proposed approach, and potentially stir more interest and effort in the related research communities and the industrial professionals.

7.1 Lessons Learned

The evolutionary topology modification and network design framework that was presented in the previous chapters has utilized various techniques already available in other disciplines of Network Science, thus combining and exploiting their benefits. As already explained, the aim was originally to eventually develop more efficient methodologies for analyzing, designing, and controlling networks, leading eventually to more efficient and useful networking structures.

In the following, we provide a brief overview of the most important emerging trends that were presented in the previous chapters and are expected to prove themselves in various similar or different studies with respect to Network Science and complex networks. Furthermore, we highlight several features of the summarized approaches that could be proven useful in future studies in various and diverse ways. We note that the corresponding list is non-exhaustive. However, it could be used as a first validation of the key issues and steps that one may take into account in similar studies and designs, or in an attempt to extrapolate such trends in other areas not yet touched from this perspective.

7.1.1 Emerging Trends and Their Benefits

Various aspects of the problems and features of the networking structures presented in the previous chapters are mostly noteworthy for future consideration and further analysis. Such emerging trends may potentially enable researchers and practitioners in modifying and/or designing well-adopted and employed complex network topologies of today more efficiently, with better control and possibly in various and diverse application frameworks. The most significant of those emerging in this book are summarized in the following:

1. Network Science is becoming one of the very promising areas of research, mainly because it brings several scientific disciplines together and it could potentially contribute towards solving more efficiently even more problems that belong in the various disciplines, by utilizing each time the more appropriate techniques (as they originally emerge in each discipline, or properly adapted to fit the special circumstances in each specific discipline).

2. The small-world paradigm has emerged as one of the very important topological structures, both in the presentation of Network Science in general, and in the specific mechanisms provided regarding topology modification of wireless multi-hop networks. It seems that the small-world effect is characteristically related to the feature of small average path length of network topologies. Ideally, a network desired to scale efficiently in terms of average network distance should employ in one way or another the small-world paradigm, or at least employ a mechanism that embeds some kind of small-world features in the network structure, which in turn will drive the scaling of the average path length towards the lower values of the allowed scale.

3. Similarly, the scale-free (power-law) regime emerging as a consequence of preferential attachment law and its variations in evolving networks has been closely tied with many desired features of networks across the disciplines of Network Science, constituting both the scale-free property and preferential attachment, two of the significant substances of networks in general. Furthermore, with respect to the preferential attachment mechanism, it has been shown how to theoretically and practically exploit it in evolutionary network modification mechanisms for improving the features and performance of certain types of networks. More importantly, a broader concept emerged, namely extending the incorporation of scale-free features and the preferential attachment mechanism into other types of network application frameworks as well.

4. Regarding evolutionary computing approaches, the process of selection emerges as a vital feature in any evolutionary algorithm. Several studies have revealed how critical selection can be for developing effective algorithms and desired performance results. For this reason, in any evolutionary process considered or designed, it is important to identify the type of selection involved and carefully considered and/or parameterized, in order to yield the desired outcomes. Similarly, crossover/mutation has a role analogous to selection, and essentially it jointly realizes with selection the "stochastic search" approach employed by all evolutionary algorithms.

5. The presented evolutionary network modification framework adopts some, but not all the features of evolutionary computing and/or Network Science methodologies. This is a broader principle to consider in the design of similar approaches, where the designer may adopt only the features that will yield the desired results, in a manner that does not complicate the operation and retains resource consumption (from any resource type) that is low, or at least to the level of the original network design. Luckily, social and complex network analysis have revealed and continue to reveal a lot of interesting features, which might become rather desired in the immediate or more distant future.

In the rest of this subsection, we focus on presenting some examples related to the aforementioned trends, in which the application of a socially-inspired topology modification mechanism in the initial wireless network yielded important benefits for the network and its users (network and application layers). These examples will further reinforce the approaches presented in the previous chapters of the book and provide additional perspective on their theoretical potential and practical applications.

To begin with, the authors in [44] discussed (mainly through simulations) the significant improvement in network connectivity from the introduction of a small number of special nodes in the network topology. These special nodes function in two radios with different transmission radii (a short and a long) serving as long-range shortcuts. A practically feasible consideration of connectivity through a percolation perspective was suggested, in order to obtain results on the information diffusion control potentials. By adding special nodes with different long-range transmission radii, it was proved that there is an upper bound on the number of special nodes that should be added and on their transmission range, to transform a disconnected topology to an almost connected topology. Thus, the power increase required for connectivity improvement is limited, while the transmission radii of only the special nodes needs to be increased for achieving connectivity with high probability. These results are closely related to some of the trends mentioned before, showing that the overall direction suggested by the identified trends have practical merit, in addition to their theoretic interest.

Another perspective of positive influence of nodes' heterogeneity on network's overall performance is the lifetime of the network, which is typically defined as the time until the first node dies due to battery outage. Towards this emerging direction, the authors in [167] studied through analysis and simulations the positive impacts of the link or energy heterogeneity on a sensor network lifetime. Regarding link heterogeneity, some nodes in the topology were assumed to possess long-range and highly reliable (with respect to success rates) links to the sink node, like shortcuts in social networks although not explicitly mentioned. Energy heterogeneity was achieved through enabling some nodes to have no energy restrictions and thus zero energy-cost for their communications, and it was analytically proved that energy heterogeneity increased the network lifetime when the high power nodes form a dominating tree routed at the sink node. Furthermore, simulation and testbed results indicated improvements in end-to-end success rates, energy consumption, and end-to-end latency. The large-scale sensor network lifetime with respect to node degree heterogeneity was studied in [104], from a percolation perspective, as the critical time before which the network keeps a giant component and the impact of the degree-dependent node lifetime upon the network lifetime is examined. Through analysis and simulations, it was demonstrated that controlling the individual node lifetime so as to exploit heterogeneity on nodes' degrees (and lifetime is especially increasing with node degree) significantly increases the overall network maintenance, a very useful guideline for designing future networks.

Other important examples with emerging useful trends and guidelines of either the application of socially-inspired topology modification mechanisms or social networks' analysis metrics include the fast consensus and decision making on small-world graphs [116], [90], [15], the random walks on small-world graphs [136], the routing and content management in wireless multi-hop networks [97], [49], etc.

7.1.2 Discussion on Evolutionary Topology Modification Mechanisms

The evolutionary topology modification and design framework presented in this book was mainly based on the notion of topology control, and it was specifically demonstrated for wireless multi-hop networks (Chapter 6), such as ad hoc, sensor, etc. These choices have mainly been made in order to facilitate the transparent and smooth demonstration of the framework, in terms not only of theory and analysis, but of implementation as well. Topology control ensures that the topology modifications made are feasible, nominal, and, in addition, can be easily implemented through simple means, i.e., power variations in this case, without further complicating the evolutionary process. The objective was to eventually demonstrate the approach rather than the complete implementation details involved. Complete details may be found in the references provided in the corresponding parts of the book.

A more careful study will reveal several aspects of the whole framework that require additional consideration, more prominently issues regarding the model extension, implementation, and application aspects of the presented framework and mechanisms. These aspects could be part of a broader study that will not only address the aforementioned concerns on realistic implementations of the presented evolutionary topology modification framework, but also reveal different mechanisms and operations for realizing similar dynamic features in other types of communications networks and possibly extrapolate the presented process to other types of networks and mechanisms by drawing the proper analogies.

7.2 The Road Ahead

The scope of this section is to collect and classify several problems from different areas of Network Science and topology modification that were presented previously, which remain open and could be of potential future value for interested researchers. Towards this direction, we first begin by providing an overview analysis of the effort accumulated until today, in order to cover the required ground towards developing the aforementioned design framework.

7.2.1 Route Covered Already

This book focused on complex network analysis metrics and complex network graphs' characteristics, as well as the development, modeling and possible applications on a wireless multi-hop network setting, aiming mainly at performance improvements. Chapter 4 described analytically all the currently used social network metrics. First, "the node degree distribution" was defined and extended to the quantity of "the node strength distribution" in weighted graphs, measuring the popularity of a node by the number of connections a node has, or by the value of the weights assigned to its connections (strength or intensity of its connections). This enabled taking into account in the analysis more than just connectivity information. The second and third metrics were the average path length of the network measuring the distance between every node pair and the clustering coefficient measuring the degree of clustering inside the network, respectively, which were also exploited in further characterizing salient features of networks, such as node distances, network robustness, coherency, and proximity. Following this, the diverse categories of centrality measures were extensively presented, allowing for the use of the most suitable centrality metrics for different applications. The most prominent metrics include the degree centrality, the closeness centrality, the betweenness centrality, the eigenvector centrality, and the subgraph centrality (Chapter 5). For directed network graphs, the centrality measure is extended to the prestige measure. Finally, another important perspective of complex networks revealed was their inherent hyperbolic structure or negative curvature, which can be used to characterize the performance of the network as a whole, which opened up a new perspective in complex and social network analysis.

Regarding the structure of social networks (Chapter 5), many of them emerge to be small-world graphs, i.e., their average path length is small and their clustering coefficient is high. Also, their degree distribution has a power-law form, characteristic of scale-free networks, showing a heterogeneity in the node degrees over the network. The Watts and Strogatz [162] model was mainly used for constructing small-world graphs, while the Barabási–Albert [6] model yields power-law graphs, both of them operating in an evolutionary constructive fashion. In general, the Watts and Strogatz model gradually introduces shortcuts in the network topology leading to path length reduction without importantly altering the clustering coefficient. The Barabási–Albert model is based on preferential attachment rules, and gradually adds both new nodes and new links enlarging the network topology, while achieving the power-law degree distribution observed in many real complex networks. Also, evolutionary models in hyperbolic space have been developed, which lead to network topologies emerging both scale-free and small-world properties. The basic intuition behind the development of evolutionary models in the hyperbolic space is the inherent hyperbolic structure existing behind many complex networks, as explained in Chapter 5. Finally, a similar property to the small-world one is the expansion property characterizing networks that are simultaneously sparse and well connected, which was described analytically in Chapter 5.

Regarding the applications of ideas from social networks analysis and evolutionary mechanisms to wireless multi-hop networks towards their improving their performance, Chapter 6 summarized all the possible existing approaches and emphasized a holistic socially-inspired topology modification mechanism. Through this holistic topology modification mechanism, we are now able, for the first time, to:

- Exploit the possible connections and similarities of wireless multi-hop with the social networks.

- Improve the network performance, while taking into account the designated mission of the network in the application layer via the network weights.

- Optimize the topology modification methodology so as to minimize the additional energy consumption.

- Provide the possibility of a distributed implementation of the topology modification mechanism when this is demanded by the network performance.

- Automate the topology modification approach towards both directions of the evolutionary design loop (top-down and bottom-up) according to incentives and various operational factors. This means that, on the one hand, we are able to enhance with shortcuts the network topology in case of a scarcely connected network with poor performance. This is achieved by assigning higher values to the parameters of the addition processes compared to the corresponding ones for the deletion processes. However, in the other case, of a densely connected topology experiencing high interference, we are able to apply SETM, parameterized towards the deletion processes leading to a sparser final topology. The latter is denoted as "inverse SETM."

The following section states and briefly discusses already identified open problems lying ahead of the aforementioned established research aspects and closely related to the research areas presented in general in this book.

7.2.2 Open Problems

The emerging problems can be classified into the following main categories. Some of them constitute straightforward extensions of the presented techniques, while others are tougher problems that require significant effort, and some could potentially prove to be too tough in the future.

- Betweenness centrality is a very interesting and important centrality measure, especially for communication networks, indicating the control that a node has in the network by participating in shortest paths. This is important in many different disciplines, such as identifying and reinforcing central nodes to avoid network disconnection, boost network efficiency, or simply identify key nodes that affect the message quality when security or QoS matters. However, the exact betweenness centrality computation currently requires centralized algorithms and high computational power, significantly increasing with the number of nodes.

Although diverse approximations of the betweenness centrality are already proposed, given the importance of betweenness centrality for the network performance, we need to invent completely distributed, light, and efficient algorithms for its computation. This will provide the possibility of designing optimized network algorithms using each node's importance, allowing for improved network performance needed for the next generation social-based communications networks.

- Another open issue is the invention of a more generic and flexible centrality definition, one that is able to cover holistically diverse and conflicting notions of node importance. An appropriate metric should also be possible to distributively compute with the minimum possible overhead with respect to the background network traffic, if any.

- In the presentation of evolutionary computation approaches, recombination and crossover emerged as significant aspects of such methodologies and in some cases, e.g., evolutionary programming, they were shown to be dominant factors, typically decisive for the operation of such techniques. Deciding the exact form and parameters of such mechanisms and their relative weight on the final outcome of an employed methodology is not straightforward, and significant research is still required in this case, especially in order to achieve this in an automated fashion as well. More specifically, the corresponding research should be more focused on the specific aspects of the evolutionary process that are of more interest within the framework of social/complex network analysis than those presented in Chapter 4 regarding the computational aspect of such algorithms.

- In Chapter 5 and Section 5.5, we briefly described the theory of expander graphs and their salient properties regarding routing and communication properties. It would be very efficient to develop mechanisms for creating wireless network topologies that were provably expander graphs, since expander graphs have sparse connections reducing the interference but are, however, well connected through short paths (similar to small-world graphs), improving the speed of communication and the retransmissions.

- Finally, another open problem is to simplify and improve the distributed implementation of SETM, so that the next generation of self-improved (alternatively self-optimized) networks is obtained, which will be able to observe their current performance and if necessary to autonomically trigger the application of SETM or inverse SETM for balancing the trade-off between performance and cost. It is important that such a procedure is realized at the node-level, without requiring complex and costly centralized mechanisms.

7.3   Epilogue

In the previous chapters, this book mainly covered subjects relevant to the dynamics of evolutionary methods for the analysis and design of wireless complex networks. The presentation adopted the framework of Network Science, which is an inherently multi-disciplinary approach, applicable to numerous and diverse branches of science, while it also employs diverse and radical methodologies from the involved scientific fields. The Network Science perspective allows, in general, the uses of complex and social network methods for modeling and studying several different types of networks, and also exploits a broad set of metrics, such as centrality and clustering coefficient, for the design, analysis, and performance assessment of the networks of interest.

Apart from the network analysis part that was mainly based on the Network Science perspective and social/complex network analysis methods, another important pillar of this book was the evolutionary design and modification of wireless complex networks. Inspired by evolutionary computing approaches, which were briefly presented, the concept of evolutionary network modification/improvement was suggested and described in a generic fashion, applicable to arbitrary types of networks, even networks regarding other scientific disciplines. However, along with the relevant objectives and goals of evolutionary network modification/improvement in the general case, some more specific application examples were provided, focusing on wireless complex networks. Specifically, this book focused on the modification and improvement of wireless ad hoc and sensor networks in different application paradigms, e.g., data flow, trust establishment, etc. It provided the general strategy and some of the more specific objectives required in order to achieve the evolutionary approach for improving the communications of interest. More specific goals and desired features were identified for the case of multi-hop networks and according to these goals and features several approaches and examples were provided demonstrating in a more concrete manner how the evolutionary modification/improvement paradigm can be applied in this specific setting of wireless ad hoc and sensor networks.

We strongly believe that the described emerging framework and its application as demonstrated in the previous chapters can be more broadly applied in other types of communications networks, by following the steps described in the specific scenarios presented and adapting them properly to conform with legitimate processes and actions in the application framework of each type of network. As already explained, the presented topics are just a small step towards more holistic frameworks, and it is expected that within the framework of Network Science similar approaches will emerge in the future, possibly in various application domains. However, it should be expected that communications networks will be dictating the pace of this paradigm shift, mainly because the corresponding trends have already been observed in various degrees in the different areas of communications networks, and because in addition, such networks can be easily modified nowadays, allowing the corresponding ideas to be rapidly verified and assessed. For instance, compared to financial or social networks, where real modifications may be costly or even impossible, not to mention the potential underlying risk, especially in networks involving biological modifications, in communications networks these modifications can be more tested and applied straightforwardly. Such aspects may be potential limiting factors for the evolution of such networks and the eventual drivers are not always necessarily dictated by logical or measurable factors, as is the case in communications networks.

It is exactly this last difference that sets communications networks and the corresponding Network Science-related methodologies in a dominant position in the broader field of network analysis and engineering. New ideas and concepts can first be developed within the communications networks framework and then research effort may be focused on drawing the proper analogies and extending these ideas to other types of networks as well. However, this means that communications networks are not expected to receive similar feedback from other disciplines in the near future, and it is for this reason that more effort should be accumulated in the communications domain in Network Science, in accordance with the guidelines and factors presented in Chapter 1. Future attempts should also focus on obtaining feedback from other neighboring fields, such as Mathematics and Physics, which have traditionally fed engineering with valuable knowledge and techniques.

Perhaps the most intriguing trend emerging from this book is that network engineering has just begun to become exciting and fascinating, and it progressively becomes an established field of its own merit. Contrary to the previous times, where networks were mainly considered a part of graph theory and thus were treated as a working knowledge that researchers would superficially acquire in order to apply it in their own fields of expertise, currently the emerging trend is that more focused scientists active in the field of Network Science will be summoned to solve emerging network problems in other fields, the same way that mathematicians might be required to solve complicated problems of mathematical nature in other disciplines.

This book attempted to contribute to the aforementioned direction, and provide some starting points for such attempts, using as application domain the field of wireless ad hoc and sensor networks. The successfulness of this goal will be measured by similar emerging attempts, both in the field of communications networks and other types of networks. Extrapolating the methodologies and results presented in this book in other disciplines of Network Science would be one of the major achievements of the current effort and we strongly believe that it would stir considerable additional research effort that would benefit not only the other disciplines of Network Science, but specifically wireless networks as well. We trust that the presented material will prove sufficient to provide the necessary background and stimulus for the aforementioned goals.

Appendices

Appendix A

Geometric Probability

This appendix serves a twofold purpose, initially presenting some basic notions from Probability Theory, which are used extensively and implicitly throughout this book, and secondly, introducing some concepts of Geometric Probability, which are used in the more focused section of Chapter 6. The latter are lying at the core of the topology modification methodology and especially at the heart of the modeling approach for the case of wireless networks demonstrated in this book. We refer only to the most basic elements of Probability Theory and Geometric Probability required in the current treatment, in order to facilitate the interested reader. For a detailed study of Probability Theory the more interested reader can refer to various established references such as [127], [64] and the more advanced and recent ones, [76], [56]. Especially for Geometric Probability three excellent references for further study are [10], [98], and [151], all of which require a more involved knowledge of probabilistic elements and tools (in an increasing order of required familiarity). A more accessible overview and working coverage of Geometric Probability can be found in [83], while [82] provides a more focused perspective of Geometric Probability from the perspective of wireless networks.

A.1 Probability Theory Elements

In the following, we first present the elements of Probability Theory required for presenting in the sequence the notions of Geometric Probability exploited and applied in various capacities in this book.

Let us begin by considering a probability space $(\Omega, \mathbb{F}, \mathbb{P})$, where Ω is the sample space consisting of all the possible outcomes, $\mathbb{F}$ is a σ-algebra, i.e., the collection of all the events (sets of outcomes of Ω) to which probabilities can be assigned, and $\mathbb{P} : \mathbb{F} \to [0, 1]$ is the probability measure (probability function) that assigns probabilities to the elements of $\mathbb{F}$. Then with respect to this probability space,

Definition 29 *A random variable is a function $X : \Omega \to \mathbb{R}$, assigning to every $\omega \in \Omega$ (i.e., to every outcome of a random experiment) the value $X(\omega) \in \mathbb{R}$.*

X is a random variable on Ω if and only if the sets $\{X(\omega) \le x\}, \forall x \in \mathbb{R}$, are events of $\mathbb{F}$, i.e., the probabilities $\mathbb{P}\{X(\omega) \le x\}, \forall x \in \mathbb{R}$ can be defined. In this case, if B is a Borel set of $\mathbb{R}$, it holds that $X^{-1}(B) \in \mathbb{F}$. If X can take only discrete values (i.e., X can only take the values $\{p_1, p_2, ..., p_n\}$) then it is a discrete-type random variable, otherwise if X can take all the values in a predefined continuous set (i.e., $X \in [a, b]$, given a, b), it is a continuous type random variable.

Example 1: In a coin tossing experiment, there are two possible outcomes, "head," denoted as "h," or "tail," denoted as "t." Thus $\Omega = \{h, t\}$. If we toss the coin only once then $\Omega \equiv \mathbb{F}$, and for a fair coin it holds that $\mathbb{P}(h) = \mathbb{P}(t) = \frac{1}{2}$. A possible discrete random variable over Ω is X, where $X(h) = 1$, $X(t) = 0$. Then $\mathbb{P}(1) = \mathbb{P}(0) = \frac{1}{2}$. In the case that the coin is tossed twice, $\Omega = \{h, t\} \times \{h, t\}$, $(F) = \{hh, ht, th, tt\}$, and a probability should be assigned to each one of the four events (which for a fair coin is assigned in an equiprobable manner).

Definition 30 *The probability distribution function of the random variable X is the function $F_X(x) = \mathbb{P}\{X \le x\}$, where $x \in [-\infty, +\infty]$.*

The distribution function has the following properties:

- $F_X(+\infty) = \mathbb{P}\{X \le +\infty\} = 1$, $F_X(-\infty) = \mathbb{P}\{X \le -\infty\} = 0$.

- $F_X(x)$ is a non-decreasing function of x, i.e., if $x_1 < x_2 \to F_X(x_1) \le F_X(x_2)$.

- $\mathbb{P}\{X > x\} = 1 - \mathbb{P}\{X \le x\}$.

- $F_X(x)$ is right continuous, i.e., $\lim_{x \to x_0^+} F_X(x) = F(x_0)$, where $x \to x_0^+$ denotes that $x \to x_0$ from the right, i.e., from higher values.

- $\mathbb{P}\{x_1 < X \le x_2\} = F_X(x_2) - F_X(x_1)$.

- $\mathbb{P}\{X = x\} = F_X(x) - F_X(x^-)$, where $F_X(x^-) = \lim_{x \to x^-} F_X(x)$.

Example 2: In Example 1, we defined the random variable X where $X(h) = 1$, $X(t) = 0$, and $\mathbb{P}(1) = \mathbb{P}(0) = \frac{1}{2}$. Regarding its distribution function, we have the following:

$$\forall \, x < 0, \; \mathbb{P}\{X < x\} = 0$$

$$\forall \, 0 \le x < 1, \; \mathbb{P}\{X < x\} = \frac{1}{2}$$

$$\forall \, x \ge 1, \; \mathbb{P}\{X < x\} = 1$$

Definition 31 *The probability density function of the random variable X, $f_X(x)$ is a non negative function defined as the derivative of the probability distribution function $F_X(x)$.*

If X is of continuous type then $f_X(x) = \frac{dF_X(x)}{dx}$, otherwise if X is of discrete type then $f_X(x) = \sum_{\forall\ i} p_i \delta(x - x_i)$, where the values x_i represent the discontinuity points of $F_X(x)$. Also, it can be obtained by integration that $F_X(x) = \int_{-\infty}^{x} f_X(u)du$.

Definition 32 *The joint probability distribution function of two random variables X and Y, denoted as $F_{XY}(x,y)$, is the probability of the event $\{X \leq x, Y \leq y\}$, where x, y are real numbers.*

The following properties hold for the joint distribution function:

- $F_{XY}(+\infty, +\infty) = 1$, $F_{XY}(-\infty, y) = 0$, $F_{XY}(x, -\infty) = 0$.

- $\mathbb{P}(\{x_1 < X \leq x_2, y_1 < Y \leq y_2\}) = F_{XY}(x_2, y_2) - F_{XY}(x_1, y_2) - F_{XY}(x_2, y_1) + F_{XY}(x_1, y_1)$

Definition 33 *The joint probability density function of two continuous type random variables X and Y, denoted as $f_{XY}(x,y)$, is defined as $f_{XY}(x,y) = \frac{\partial^2 F_{XY}(x,y)}{\partial x \partial y}$.*

Definition 34 *Two random variables X, Y are called independent if the events $\{X \in A\}$, $\{Y \in B\}$ for all Borel sets A, B, are independent, i.e., if $\mathbb{P}(\{X \in A, Y \in B\}) = \mathbb{P}\{X \in A\}\mathbb{P}\{Y \in B\}$ and as a result, $f_{XY}(x,y) = f_X(x)f_Y(y)$.*

A.2   Probabilistic Modeling of the Deployment of a Wireless Multi-Hop Network

Wireless multi-hop networks are typically modeled as Random Geometric Graphs (RGGs), as already mentioned in Chapter 1 (Section 1.2.2). In turn, the RGG model of a wireless multihop network can be regarded as a spatial point process. A spatial point process describes a random pattern of points/nodes distributed over a space of given dimensions [10]. For the case of wireless multihop it suffices to assume a plane square region A of side length L as the space over which the random pattern of points/nodes is deployed. We consider N points distributed uniformly and randomly over A. In this arrangement, the coordinates of points/nodes (x, y) in the area A can be perceived as the values of two independent random variables X, Y, due to the randomness in the way the points were distributed over the plane A [98]. Then, assuming the point process is simple (i.e., with probability 1, no two points are coincidental) and locally finite (i.e., any bounded region contains only a finite or countably infinite number of points with probability 1), the probability of the event {a point (x, y) is located in a subarea of region L^2}, is described by the joint density function of X, Y:

$$f_{XY}(x, y) = f_X(x)f_Y(y) = \begin{cases} \frac{1}{L^2}, & (x, y) \in A \\ 0, & \text{otherwise} \end{cases} \tag{A.1}$$

provided that X, Y are independent random variables uniformly distributed in $[0, L]$ [10].

Similarly, the probability that a point (x, y) belongs to the subarea B of the area L^2 equals:

$$\mathbb{P}\{(x, y) \in B\} = \int \int_B f_{XY}(x, y)dxdy = \frac{|B|}{L^2} \tag{A.2}$$

where $|\cdot|$ denotes the measure of a set.

The above means that this probability depends only on the size of the geometric subarea of region L^2 considered, and it is identical to the probability that the point (x, y) belongs to another subarea C of the region of the same measure $|C| = |B|$. This is due to the independence of X, Y, and it could be considered a form of spatial Markov property , in a manner similar to the time expression of Markov property (memoryless property).

Appendix B

Semirings and Path Problems

In this appendix two very important notions, both for the broader field of shortest path problems on graphs (discrete optimization), as well as the topology modification approaches presented in this book, are presented. More specifically, in the current setting, these two structures are used for solving the generalized shortest path problem in weighted graphs. Two such algebraic structures are the monoid and the semiring, where the definition of the latter is based on the definition of the former.

The notion of semiring is widely used for the modeling and computation of the generalized shortest paths in weighted graphs spanning various and diverse application frameworks. The semiring can be adapted via its two operators $\oplus, \otimes$ for expressing diverse notions of "shortest paths," as is will be explained shortly. In the following, we first present the definitions of the monoid and the semiring, and then illustrate their use for the computation of the generalized shortest paths through some indicative examples.

B.1 Monoids

A monoid is a set S, together with a binary operation $\odot$, denotes as $(S, \odot)$, which satisfies the following three axioms:

- Closure: For all a, $b \in S$, also $a \odot b \in S$.

- Associativity: For all a, b and $c \in S$, the equation $(a \odot b) \odot c = a \odot (b \odot c)$ holds.

- Identity element: There exists an element $e \in S$, such that $\forall a \in S$, the equation $e \odot a = a \odot e = a$ holds.

If $a \odot b = b \odot a$, i.e., the operator $\odot$ is commutative, then the monoid is characterized as commutative monoid.

B.2   Semirings

A semiring is an algebraic structure $(S, \oplus, \otimes)$, where S is a set and $\oplus, \otimes$ are binary operators with the following properties:

- $\oplus$ is commutative, associative with neutral element $\hat{0} \in S$, which means that $(S, \oplus)$ is a commutative monoid, i.e., for a, b, $c \in S$:

$$a \oplus b = b \oplus a$$

$$(a \oplus b) \oplus c = a \oplus (b \oplus c)$$

$$a \oplus \hat{0} = a$$

- $\otimes$ is associative with neutral element $\hat{1} \in S$ and absorbing element $\hat{0} \in S$, which means that $(S, \otimes)$ is a monoid, i.e., for a, b, $c \in S$:

$$(a \otimes b) \otimes c = a \otimes (b \otimes c)$$

$$a \otimes \hat{1} = \hat{1} \otimes a = a$$

$$a \otimes \hat{0} = \hat{0} \otimes a = \hat{0}$$

- $\otimes$ distributes over $\oplus$:

$$(a \oplus b) \otimes c = (a \otimes c) \oplus (b \otimes c)$$

$$a \otimes (b \oplus c) = (a \otimes b) \oplus (a \otimes c)$$

The operators of the semiring can be applied in the case of computing the generalized shortest path problem, in the following sense. The operator $\otimes$ is used to combine the values of the weights along a single path, while the operator $\oplus$ serves the comparison among the values of different paths (comparison of path length across paths) and the choice of the "shortest" one. Let us consider the path $p = (v_0, v_1, v_2, ..., v_n)$ starting from node v_0 and ending at v_n, where v_i, $i = 0, 1, ..., m$ are the network vertices and $W = [w(v_i, v_j)]$ is the weight matrix. Then the generalized weight of the path p is equal to $w_p = w(v_0, v_1) \otimes w(v_1, v_2) \otimes ... \otimes w(v_{n-1}, v_n)$. Let us consider that there are $l > 1$ paths connecting the nodes v_0, v_n. Then the generalized shortest path will have a weight equal to $d(v_0, v_n) = w_{p1} \oplus w_{p2} \oplus ... w_{pl}$. To illustrate this computation, the generalized shortest path between nodes i, j in Fig. B.1 (the values on the links represent their corresponding weights) equals to $d(i, j) = \left((w_{ik} \otimes w_{kl}) \oplus (w_{im} \otimes w_{ml})\right) \otimes w_{ln} \otimes w_{nj}$.

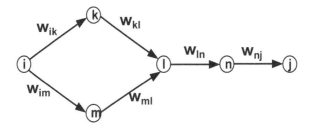

Figure B.1: Network graph for the computation of the generalized shortest path problem.

B.3 Examples

In the following, we present three specific examples of generalized shortest path formulations/computations via semirings.

In the first, in order to compute the minimum cost path between each pair of nodes (i.e., the weights represent cost values), the semiring $S_1 = (\Re \cup \infty, \min, +)$ is employed. The weight of each path is computed by summing the cost of its links, while the path chosen is the one with the minimum cost value.

In the second example, if one desires to compute the maximum possible rate of traffic between each pair of nodes (the weights of the links represent the maximum rate that can be supported by the corresponding link), the semiring $S_2 = (\Re \cup \infty, \max, \min)$ can be employed. This is so because the maximum rate that can be supported by each path is the minimum rate of all its links, while the path that is able to support the maximum rate is chosen.

Finally, in the case that the weights of network edges model trust values that are defined in the set $[0, 1]$ (with zero representing no trust at all, while the unity representing total trust) the suitable semiring to be employed is $S_3 = ([0, 1], \max, \times)$. This semiring can be interpreted in the following manner. The weight/trust value of a path is the product of the trust values of its constituent links and the trust value that node v_i assigns to node v_j (if they are not directly connected) equals to the maximum trust value of all the paths beginning from v_i and ending at v_j.

For a detailed study of the path problems of networks based on semirings, the interested reader should refer to [16], [112], [22].

References

[1] R. Albert and A.-L. Barabási. Emergence of scaling in random networks. *Science*, 401, 1999.

[2] L. Adamic. The small world web. *Lecture Notes in Computer Science*, 1696, 1999.

[3] Y. Afek and Y. Matias. Elections in anonymous networks. *Journal of Information and Computation*, 113, 1994.

[4] N. Afifi and K.-S. Chung. Small world wireless mesh networks. In *International Conference on Innovations in Information Technology (IIT)*, 2008.

[5] R. Albert and A.-L. Barabási. Topology of evolving networks: Local events and universality. *Physical Review Letters*, 85(24):5234–5237, December 2000.

[6] R. Albert and A.-L. Barabási. Statistical mechanics of complex networks. *Reviews of Modern Physics*, 74(1):47–97, January 2002.

[7] R. Albert, H. Jeong, and A.-L. Barabási. Diameter of the world wide web. *Nature*, 401, 1999.

[8] J. W. Anderson. *Hyperbolic Geometry*, 2nd Edition, Springer, London, UK, 2007.

[9] T. Back, D. B. Fogel, and Z. Mickalewicz (Eds.). *Evolutionary Computation 1: Basic Algorithms and Operators*. Institute of Physics Publishing, Bristol, U.K., 2000.

[10] A. Baddeley, I. Barany, R. Schneider, and W. Weil. *Stochastic Geometry*. Lecture Notes in Mathematics, Springer-Verlag, Berlin-Heidelberg, Germany, 2004.

[11] D. A. Bader, S. Kintali, L. Madduri, and M. Mihail. Approximating betweenness centrality. *5th International Conference on Algorithms and Models for the Web-Graph (WAW)*, San Diego, CA, U.S.A., 2007.

[12] A.-L. Barabási. Scale-free networks: A decade and beyond. *Science*, 325(5939):412–413, July 2009.

[13] A.-L. Barabási and E. Bonabeau. Scale-free networks. *Scientific American*, 288(May):2003.

[14] A.-L. Barabási, H. Jeong, Z. Neda, E. Ravasz, A. Schubert, and T. Vicsek. Evolution of the social network of scientific collaborations. *Physica A: Statistical Mechanics and its Applications*, 311(3), 2002.

[15] J. S. Baras and P. Hovareshti. Efficient and robust communication topologies for distributed decision making in networked systems, pages 3751–3756, December 2009.

[16] J. S. Baras and G. Theodorakopoulos. *Path Problems in Networks*. Synthesis Lectures on Communication Networks, Morgan and Claypool Publishers, 2010.

[17] A. Barrat, M. Barthélemy, R. Pastor-Satorras, and A. Vespignani. The architecture of complex weighted networks. *PNAS*, 101(11):3747–3752, March 2004.

[18] A. Barrat, M. Barthélemy, and A. Vespignani. Modeling the evolution of weighted networks. *Physical Review Letters*, 70(6), December 2004.

[19] A. Barrat, M. Barthélemy, and A. Vespignani. Traffic-driven model of the world wide web graph. *Lecture Notes in Computer Science*, 3243, 2004.

[20] A. Barrat, M. Barthélemy, and A. Vespignani. Weighted evolving networks: Coupling topology and weight dynamics. *Physical Review Letters*, 92(22), June 2004.

[21] M. Barthélemy, A. Barrat, R. Pastor-Satorras, and A. Vespignani. Characterization and modeling of the weighted networks. *Physica A, Elsevier*, 346, 2005.

[22] V. Batagelj. Semirings for social networks analysis. *Journal of Mathematical Sociology*, 19(1), 1994.

[23] A. Bavelas. Communication patterns in task oriented groups. *Journal of the Acoustical Society of America*, 22, 1950.

[24] M. A. Beauchamp. An improved index of centrality. *Behavioural Science*, 10(2):161–163, 1965.

[25] D. P. Bertsekas. *Non Linear Programming*. Athena Scientific, Nashua, WH, U.S.A., 1999.

[26] D. P. Bertsekas and R. G. Gallager. *Data Networks*. Prentice Hall, Englewood Cliffs, NJ, U.S.A., 1992.

[27] D. P. Bertsekas, A. Nedic, and A. E. Ozdaglar. *Convex Analysis and Optimization*. Athena Scientific, Nashua, NH, U.S.A.., 2003.

[28] D. Bertsimas and J. Tsitsiklis. Simulated annealing. *Statistical Science*, 8(1), 1993.

[29] G. Bianconi and A.-L. Barabási. Bose-einstein condensation in complex networks. *Physical Review Letters*, 86(24), June 2001.

[30] G. Bianconi and A.-L. Barabási. Competition and multiscaling in evolving networks. *Europhysics letters*, 54(4):436–442, May 2001.

[31] D. M. Blough, M. Leoncini, G. Resta, and P. Santi. The k-neighbors approach to interference bounded and symmetric topology control in ad hoc networks. *IEEE Transactions on Mobile Computing*, 5(9):1267–1282, September 2006.

[32] M. Boguná, F. Papadopoulos, and D. Krioukov. Sustaining the internet with hyperbolic mapping. *Nature*, 1(62), September 2010.

[33] B. Bollobas. *Modern Graph Theory*. Springer Verlag, New York, U.S.A.,

1998.

[34] B. Bollobas and O. Riordan. *Percolation*. Cambridge University Press, Cambridge, U.K., 2006.

[35] P. Bonacich. Factoring and weighting approaches to status scores and clique identification. *Journal of Mathematical Sociology*, 2:113–120, 1972.

[36] P. Bonacich. Power centrality: A family of measures. *The American Journal of Sociology*, 92(5):1170–1182, 1987.

[37] A. Bonato. *A Course on the Web Graph, Graduate Studies in Mathematics*. American Mathematical Society, Providence, RI, U.S.A., Vol. 89, 2008.

[38] S. P. Borgatti. Centrality and network flow. *Social Networks (Elsevier)*, pages 55–71, 2004.

[39] S. Boyd and L. Vandenberghe. *Convex Optimization*. Cambridge University Press, Cambridge, U.K., 2004.

[40] U. Brandes. A faster algorithm for betweenness centrality. *Journal of Mathematical Sociology*, 25(2), 2001.

[41] U. Brandes and C. Pich. Centrality estimation in large networks. *International Journal of Bifurcation and Chaos: Special Issue on Complex Networks' Structure and Dynamics*, 34, 2007.

[42] I. Broustis, G. Jakllari, T. Repantis, and M. Molle. A comprehensive comparison of routing protocols for large-scale wireless manets. *3rd Annual IEEE Communications Society on Sensor and Ad Hoc Communications and Networks (SECON)*, 3, 2006.

[43] C. C. Chiang, H.-K. Wu, W. Liu, and M. Gerla. Routing in clustered multihop, mobile wireless networks with fading channel. *6th IEEE International Conference on Universal Personal Communications Record*, 2, 1997.

[44] D. Cavalcanti, D. Agrawal, J. Kelner, and D. Sadok. Exploiting the small-world effect to increase connectivity in wireless ad hoc networks. In *Telecommunications and Networking — ICT*, LCNS 3124, Fortaleza, Brazil, pp. 388–393, August 2004.

[45] I. Chavel. *Riemannian Geometry: A Modern Introduction*. New York: Cambridge University Press, 1994.

[46] M. Chiang, S. H. Low, A. R. Calderbank, and J. C. Doyle. Layering as optimization decomposition: A mathematical theory of network architectures. *Proceedings of the IEEE*, 95(1):255–312, September 2007.

[47] R. Chitradurga and A. Helmy. Analysis of wired short-cuts in wireless sensor networks. In *IEEE/ACS International Conference on Pervasive Services*, Beirut, Lebanon, July 2004.

[48] T. H. Cormen, C. E. Leiserson, and R. L. Rivest. *Introduction to Algorithms*. MIT Press and McGraw-Hill, Third Edition, Cambridge, MA, U.S.A., 2009.

[49] E. M. Daly and M. Haahr. Social network analysis for information flow in disconnected delay-tolerant manets. *IEEE Transactions on Mobile*

Computing, 8(5):606–621, May 2009.

[50] R. Diestel. *Graph Theory.* Graduate Texts in Mathematics 173 (Springer), Heidelberg, Germany, 2005.

[51] E. W. Dijkstra. A note on two problems in connection with graphs. *Numerische Mathematik*, 1, 1959.

[52] P. S. Dodds, R. Muhamad, and D. J. Watts. An experimental study of the search in global social networks. *Science*, 301(5634):827–829, August 2003.

[53] M. Dorigo, M. Birattari, and T. Stutzle. Ant colony optimization. *IEEE Computational Intelligence Magazine*, 1(4):28–39, November 2006.

[54] S. N. Dorogovtsev, J. F. F. Mendes, and A.N. Samukhin. Structure of growing networks: Exact solution of the barabási-albert model. *Physical Review Letters*, 85, 2000.

[55] S. N. Dorogovtsev and J.F.F. Mendes. Exactly solvable analogy of small-world networks. *Europhysics Letters*, 50, 2000.

[56] R. Durrett. *Probability: Theory and Examples (3rd Ed.).*

[57] D. Easley and J. Kleinberg. *Networks, Crowds and Markets: Reasoning about a Highly Connected World.* Cambridge University Press, 2010.

[58] The Economist. A market for computing power. *The Economist Newspaper Limited*, 2011.

[59] A. E. Eiben and J. E. Smith. *Introduction to Evolutionary Computing.* Springer-Verlag, Berlin, Germany, 2003.

[60] D. Eppstein and J. Wang. Fast approximation of centrality. *Journal of Graphs Algorithms and Applications*, 8, 2004.

[61] E. Estrada. Spectral scaling and good expansion properties in complex networks. *Europhysics Letters*, 73, 2006.

[62] E. Estrada and J. A. Rodríguez-Velázquez. Subgraph centrality in complex networks. *Physical Review Letters*, 71, 2005.

[63] G. Fagiolo. Clustering in complex directed networks. *Physical Review Letters*, 76(2), August 2007.

[64] W. Feller. *An Introduction to Probability Theory and its Applications (3rd Ed.).*

[65] D. B. Fogel. *Evolutionary Computation.* IEEE Press, Piscataway, NJ, U.S.A., 1995.

[66] L. J. Fogel, A. J. Owens, and M. J. Walsh. *Artificial Intelligence through Simulated Evolution.* Wiley, Chichester, U.K., 1966.

[67] M. L. Fredman and R. E. Tarjan. Fibonacci heaps and their uses in improved network optimization algorithms. *Journal for the Association of Computing Machinery*, 34, 1987.

[68] L. Freeman. A set of measures of centrality based on betweenness. *Sociometry*, 40(1), 1977.

[69] L. C. Freeman. Centrality in social networks conceptual clarification. *Social Networks*, 1, 1978–79.

[70] L. C. Freeman, S. Borgatti, and D. R. White. Centrality in valued graphs: A measure of betweenness based on network flow. *Social Net-*

works, 13, 1991.

[71] N. E. Friedkin. Theoretical foundations for centrality measures. *The American Journal of Sociology*, 96(6):1478–1504, May 1991.

[72] A. Fronczak, P. Fronczak, and J. A. Holyst. Average path length in random networks. *Physical Review Letters*, 70(5), November 2004.

[73] R. Geisberger, P. Sanders, and D. Schultes. Better approximation of betweenness centrality. *Journal of the Acoustical Society of America*, 1950.

[74] C. Gkantsidis, M. Mihail, and A. Saberi. Conductance and congestion in power law graphs. In *ACM SIGMETRICS*, 2003.

[75] D. Goodman. The wireless internet: Promises and challenges. *IEEE Computer*, 33(7):36–41, July 2000.

[76] G. Grimmett and D. Stirzaker. *Probability and Random Processes (3rd Ed.)*.

[77] G. R. Grimmett. *Percolation (Grundlehren der mathematischen Wissenschaften)*. Springer Verlag, Berlin, Germany, 1999.

[78] D. L. Guidoni, R. A. F. Mini, and A. A. F. Loureiro. On the design of resilient heterogeneous wireless sensor networks based on small world concepts. *Computer Networks*, 54(8):1266–1281, June 2010.

[79] R. Guimerà, S. Mossa, A. Turtsci, and L. A. N. Amaral. The worldwide air transportation network: Anomalous centrality, community structure, and cities' global roles. *PNAS*, 102(22), 2005.

[80] X. Guoliang, L. Chenyang, R. Pless, and H. Qingfeng. Impact of sensing coverage on greedy geographic routing algorithms. *IEEE Transactions on Parallel and Distributed Systems*, 17(4):348–360, April 2006.

[81] P. Gupta and P. R. Kumar. The capacity of wireless networks. *IEEE Transactions on Information Theory*, 46(2):388–404, March 2000.

[82] M. Haenggi. *Stochastic Geometry for Wireless Networks*, Cambridge University Press, Cambridge, UK, 2005.

[83] M. Haenggi, J.G. Andrews, F. Baccelli, O. Dousse, and M. Franceschetti. Stochastic geometry and random graphs for the analysis and design of wireless networks. *IEEE Journal on Selected Areas in Communications (JSAC)*, 27(7):1029–1046, September 2009.

[84] F. Harary. *Graph Tehory*. Addison-Wesley, Reading, MA, U.S.A., 1969.

[85] D. He, X. Rui, L. Weixuan, and T. G. Habetler. A survey on consumer electronic products in residential home for demand response. *38th Annual Conference on IEEE Industrial Electronics Society (IECON)*, Montreal, Canada, 2012.

[86] A. Helmy. Small worlds in wireless networks. *IEEE Communications Letters*, 7(10):490–492, October 2003.

[87] J. H. Holland. Genetic algorithms and the optimal allocation of trials. *SIAM Journal of Computing*, 2, 1973.

[88] P. Holme, S. M. Park, B. J. Kim, and C. R. Edling. Korean university life from a network perspective: Dynamics of a large affiliation network. *Physica A: Statistical Mechanics and its Applications*, 373(1):821–830,

January 2007.

[89] S. Hoory, N. Linial, and A. Widgerson. Expander graphs and their applications. *American Mathematical Society*, 43(4):439–561, October 2006.

[90] P. Hovareshti, J. S. Baras, and V. Gupta. Average consensus over small world networks: A probabilistic framework. In *47th IEEE Conference on Decision and Control*, December 2008.

[91] Y. J. Hwang, S.-W. Ko, S. I. Lee, B. I. Cho, and S.-L. Kim. Wireless small-world networks with beamforming. In *IEEE Region 10 Conference TENCON*, Fukuoka, Japan, 2009.

[92] J. A. Thorpe and I. M. Singer. *Lecture Notes on Elementary Topology and Geometry*. New York: Springer-Verlag, 1996.

[93] M. O. Jackson. *Social and Economic Networks*. Princeton University Press, Princeton, NJ, U.S.A., 2008.

[94] C.-J. Jiang, C. Chen, J.-W. Chang, R.-H. Jan, and T. C. Chiang. Construct small worlds in wireless networks using data mules. *IEEE International Conference on Sensor Networks, Ubiquitous and Trustworthy Computing (SUTC)*, Taichung, Taiwan.

[95] S. D. Kamvar, M. T. Schlosser, and H. Garcia-Molina. The eigentrust algorithm for reputation management in p2p networks. In *12th International WWW Conference*, Budapest, Hungry, 2003.

[96] V. Karyotis and S. Papavassiliou. Topology control in cooperative ad hoc networks. Chapter in *Cooperative Wireless Communications*, CRC Press, Taylor & Francis Group, Editors: Y. Zhang, H.-H. Chen, and M. Guizani, Boca Raton, FL, U.S.A., 2009.

[97] D. Katsaros, N. Dimokas, and L. Tassiulas. Social network analysis concepts in the design of wireless ad hoc network protocols. *IEEE Network*, 24(6):23–29, December 2010.

[98] D. A. Klain and G.-C. Rota. *Introduction to Geometric Probability*. Cambridge University Press, New York, NY, U.S.A., 1997.

[99] J. Kleinberg. The small-world phenomenon: An algorithmic perspective. In *32nd ACM Symposium on Theory of Computing*, Portland, OR, U.S.A., 2000.

[100] P. L. Krapivsky, S. Redner, and F. Leyvraz. Connectivity of growing random networks. *Physical Review Letters*, 85(21):4629–4632, November 2000.

[101] D. Krioukov, F. Papadopoulos, M. Kitsak, A. Vahdat, and M. Boguná. Hyperbolic geometry of complex networks. *Physical Review Letters*, 82(3), September 2010.

[102] R. V. Kulkarni, E. Almaas, and D. Stroud. Exact results and scaling properties of small-world networks. *Physical Review Letters*, 61(4), April 2000.

[103] Andrea Landherr, Bettina Friedl, and Julia Heidemann. A critical review of centrality measures in social networks. *Business & Information Systems Engineering*, pages 371–385, 2010.

[104] C.-H. Lee and D. Y. Eun. Exploiting heterogeneity to prolong the lifetime of large-scale wireless sensor networks. In *IEEE International Conference on Communications (ICC)*, Kyota, Japan, 2011.

[105] E. A. Lee. Cyber physical systems: Design challenges. *11th IEEE International Symposium on Object Oriented Real-Time Distributed Computing (ISORC)*, Orlando, FL, U.S.A., 2008.

[106] S. Lin and B. Kernighan. An effective heuristic algorithm for the traveling salesman problem. *Operations Research*, 21, 1973.

[107] B. Liu, Z. Liu, and D. Towsley. On the capacity of hybrid wireless networks. In *IEEE INFOCOM*, volume 2, 2003.

[108] S. Lui, S. Gopalakrishnan, L. Xue, and W. Qixin. Cyber-physical systems: A new frontier. *IEEE International Conference on Sensor Networks, Ubiquitous and Trustworthy Computing (SUTC)*, Taichung, Taiwan, 2008.

[109] W. G. Lycan. *Mind and Cognition: An Anthology (2nd edition)*. Blackwell Publishers, Malden, MA, U.S.A., 1999.

[110] F. D. Malliaros and V. Megalooikonomou. Expansion properties of large social graphs. In *16th International Conference on Database Systems for Advanced Applications (DASFAA'11)*, Hong Kong, China, 2011.

[111] C. P. Mayer and O. P. Waldhorst. Offloading infrastructure using delay tolerant networks and assurance of delivery. *IFIP Wireless Days (WD)*, 2011.

[112] M. Mohri. Semiring frameworks and algorithms for shortest-distance problems. *Journal of Automata, Languages and Combinatorics*, 7(3):321–350, January 2002.

[113] D. Morton. Consumer electronics: The last fifteen years. *IEEE Circuits and Devices Magazine*, 15(3):28–30, May 1999.

[114] O. Narayan and I. Saniee. The large scale curvature of networks. *Physical Review Letters*, 84(6):1539–3755, December 2011.

[115] M. Newman. The structure and function of complex networks. *SIAM Review*, 45, 2003.

[116] R. Olfati-Saber. Ultrafast consensus in small-world networks. In *American Control Conference*, Portland, OR, U.S.A., volume 4, pages 2371–2378, June 2005.

[117] Committee on Network Science for Future Army Applications. *Network Science*. National Research Council, Washington, D.C., U.S.A., 2000.

[118] J.-P. Onnela, J. Saramäki, J. Kertész, and K. Kaski. Intensity and coherence of motifs in weighted complex networks. *Physical Review Letters*, 71(6), May 2005.

[119] T. Opsahl, F. Agneessens, and J. Skvoretz. Node centrality in weighted networks: Generalizing degree and shortest paths. *Social Networks*, 32:245–251, 2010.

[120] T. Opsahl and P. Panzarasa. Clustering in weighted networks. *Social Networks*, 31(2), 155–163, 2009.

[121] L. Page and S. Brin. The pagerank citation ranking: Bringing order to

the web. *Report in Stanford Digital Library Technologies Project*, 1998.

[122] P. Pantazopoulos, M. Karaliopoulos, and I. Stavrakakis. Centrality-driven scalable service migration. *23rd Intl Teletraffic Congress(ITC)*, San Francisco, CA, U.S.A., 2011.

[123] F. Papadopoulos, M. Kitsak, M. Serrano, M. Boguná, and D. Krioukov. Popularity versus similarity in growing networks. *Nature*, 489, September 2012.

[124] F. Papadopoulos, D. Krioukov, M. Boguná, and A. Vahdat. Greedy forwarding in dynamic scale-free networks embedded in hyperbolic metric spaces. In *IEEE INFOCOM*, 2010.

[125] F. Papadopoulos, D. Krioukov, A. Vahdat, and M. Boguná. Curvature and temperature of complex networks. *Physical Review Letters*, 80(3), September 2009.

[126] S. Papavassiliou and J. Zhou. A continuum theory-based approach to the modeling of dynamic wireless sensor networks. *IEEE Communications Letters*, 9(4):337–339, April 2005.

[127] A. Papoulis and S. U. Pillai. *Probability, Random Variables and Stochastic Processes*. McGraw-Hill, New York, NY, U.S.A., 2002.

[128] M. Penrose. *Random Geometric Graphs*. Oxford University Press, Cambridge, U.K., 2003.

[129] N. Perra and S. Fortunato. Spectral centrality measures in complex networks. *Physical Review Letters*, 78(3), September 2008.

[130] S. U. Pillai, T. Suel, and S. Cha. The perron-forbenius theorem. *IEEE Signal Processing Magazine*, 22(2):62–75, March 2005.

[131] The Erdős Number Project. http://www.oakland.edu/enp/. Oakland University, Rochester, MI, U.S.A.

[132] I. Rechenberg. *Optimierung Technischer Systeme nach Prinzipien des Biologischen Evolution*. Fromman-Holzboog Verlag, Stuttgart, Germany, 1973.

[133] A. Reddy, S. Shakkottai, and L. Ying. Distributed power control in wireless ad hoc networks using message passing: Throughput optimality and network utility maximization. In *42nd Annual Conference on Information Sciences and Systems (CISS)*, 2008.

[134] S. Redner. How popular is your paper? An empirical study of the citation distribution. *The European Physical Journal*, 2, 1998.

[135] J. Rennie and G. Zorpette. The social era of the web starts now. *IEEE Spectrum*, 48(6):30–33, June 2003.

[136] B. A. Rezaei, N. Sarshar, and V. P. Roychowdhury. Random walks in a dynamic small-world space: robust routing in large-scale sensor networks. In *60th IEEE Vehicular Technology Conference*, volume 7, pages 4640–4644, December 2004.

[137] A. Reznik, S. R. Kulkarni, and S. Verdú. A small world approach to heterogeneous networks. *Communication in Information and Systems*, 3(4):325–348, September 2004.

[138] B. P. Rimal, C. Eunmi, and I. Lumb. A virtualization infrastructure

that supports pervasive computing. *5th International Joint Conference on INC, IMS and IDC (NCM)*, 2009.

[139] J. J. Rotman. *An Introduction to the Theory of Groups*. Graduate texts in mathematics 148 (Springer), Berlin, Germany, 1995.

[140] E. M. Royer and T. Chai-Keong. A review of current routing protocols for ad hoc mobile wireless networks. *IEEE Personal Communications*, 6(2):46–55, April 1999.

[141] L. Rudolph. A virtualization infrastructure that supports pervasive computing. *IEEE Pervasive Computing*, 8(4):8–13, October-December 2009.

[142] P. Santi. *Topology control in wireless ad hoc and sensor networks*. John Wiley & Sons, West Sussex, UK, 2005.

[143] J. Saramäki, M. Kivelä, J.-P. Onnela, K. Kaski, and J. Kertész. Generalizations of the clustering coefficient to weighted complex networks. *Physical Review Letters*, 75(2), February 2007.

[144] P.-H. Schwefel. *Evolution and Optimum Seeking*. Wiley, New York, NY, U.S.A., 1995.

[145] G. Sharma and R. Mazumdar. Hybrid sensor networks: A small world. In *6th ACM International Symposium on Mobile Ad hoc Networking and Computing (MobiHoc'05)*, Urbana, IL, U.S.A., 2005.

[146] G. Sharma and R. Mazumdar. A case for hybrid sensor networks. *IEEE/ACM Transactions on Networking*, 16(5):1121–1132, October 2008.

[147] E. Stai, V. Karyotis, and S. Papavassiliou. A socially-driven topology improvement framework with applications in content distribution and trust management. *Journal of Internet Services and Applications, Springer*, 2(2):113–127, June 2011.

[148] E. Stai, V. Karyotis, and S. Papavassiliou. Topology enhancements in wireless multi-hop networks: A top-down approach. *IEEE Transactions on Parallel and Distributed Systems*, 23(7):1344–1357, July 2012.

[149] E. Stai, V. Karyotis, and S. Papavassiliou. Wireless multi-hop network topology control optimization and trade-off analysis. In *35th IEEE Sarnoff Symposium (SARNOFF)*, Jersey City, NJ, U.S.A., pages 1–6, May 2012.

[150] D. Stauffer and A. Aharony. *Introduction To Percolation Theory*. Taylor and Francis Group, Philadelphia, U.S.A., 2003.

[151] D. Stoyan, W.S. Kendall, and J. Mecke. *Stochastic Geometry and its Applications (2nd Ed.)*, John Wiley & Sons, West Sussex, UK.

[152] S. H. Strogatz. *Sync: How Order Emerges From Chaos In the Universe, Nature, and Daily Life*. Hyperion, New York, NY, U.S.A., 2004.

[153] G. Theodorakopoulos and J. S. Baras. On trust models and trust evaluation metrics for ad hoc networks. *IEEE JSAC*, 24(2):318–328, February 2006.

[154] J. Travers and S. Milgram. An experimental study of the small world problem. *Sociometry*, 32(4):425–443, December 1969.

[155] V. A. Trubin. Strength of a graph and packing of trees and branchings. *Cybernetics and Systems Analysis*, 29, 1993.

[156] D. Tse and P. Viswanath. *Fundamentals of Wireless Communications.* Cambridge University Press, Cambridge, UK, 2005.

[157] Hyperbolic Toolbox v1.0. http://egl.math.umd.edu/software.html. Experimental Geometry Lab, University of Maryland, College Park, MD, U.S.A.

[158] C. K. Verma, B. R. Tamma, B. S. Manoj, and R. Rao. A realistic small-world model for wireless mesh networks. *Communication in Information and Systems*, 15(4):455–457, April 2011.

[159] H.-M. Voigt, W. Ebeling, I. Rechenberg, and H.-P. Schwefel (Eds.). *4th Conference on Parallel Problem Solving from Nature.* Springer Lecture Notes in Computer Science, Berlin, Germany, 1996.

[160] W.-X. Wang, B. Hu, T. Zhou, B.-H. Wang, and Y.-B. Xie. Mutual selection model for weighted networks. *Physical Review Letters*, 72, October 2005.

[161] R. Wattenhofer and A. Zollinger. Xtc: A practical topology control algorithm for ad-hoc networks. In *4th International IEEE Workshop on Algorithms for Wireless, Mobile, Ad Hoc and Sensor Networks (WMAN)*, Nice, France, April 2003.

[162] D. J. Watts. *Small Worlds: The Dynamics of Networks between Order and Randomness.* Princeton Studies in Complexity, Princeton University Press, Princeton, NJ, U.S.A., 1999.

[163] D. J. Watts and S. H. Strogatz. Collective dynamics of 'small-world' networks. *Nature*, 393:440–442, June 1998.

[164] J. E. Wieselthier, G. D. Nguyen, and A. Ephremides. Energy-efficient broadcast and multicast trees in wireless networks. *Mobile Networks and Applications*, 7(6):481–492, December 2002.

[165] R. A. Wilson. *The Finite Simple Groups.* Graduate Texts in Mathematics 251 (Springer-Verlag), Berlin, Germany, 2009.

[166] B. Xing, M. Deshpande, S. Mehrotta, and N. Venkatasubramanian. Gateway designation for timely communications in instant mesh networks. *8th IEEE International Conference on Pervasive Computing and Communications Workshops (PERCOM Workshops)*, Manheim, Germany, 2010.

[167] M. Yarvis, N. Kushafnagar, H. Singh, A. Rangarajan, Y. Lin, and S. Singh. Exploiting heterogeneity in sensor networks. In *IEEE INFOCOM*, volume 2, March 2005.

[168] S. H. Yook, H. Jeong, A.-L. Barabási, and Y. Tu. Weighted evolving networks. *Physical Review Letters*, 86(25), June 2001.

[169] B. Zhang and S. Horvath. A general framework for weighted gene co-expression network analysis. *Statistical Applications in Genetics and Molecular Biology*, 4(1), August 2005.

[170] H. V. Zhao, W. S. Lin, and K. J. R. Liu. *Behavior Dynamics in Media-Sharing Social Networks.* Cambridge University Press, Cambridge, U.K., 2011.

Author Index

A

Adamic, L., 156, 157, 222
Afek, Y., 185
Afifi, N., 203, 208, 222
Agneessens, F., 124, 126, 128
Agrawal, D., 249
Aharony, A., 45
Albert, R., 5, 142, 156, 157, 159,
        161, 163, 165, 166, 167,
        171
Almaas, E., 147
Amaral, L. A. N., 162, 185
Anderson, J. W., 177
Andrews, 259

B

Baccelli, F., 259
Back, T., 74
Baddeley, A., 259, 261, 262
Bader, D. A., 127, 128, 131, 134
Barabási, A.-L., 5, 142, 149, 156,
        157, 159, 160, 161, 163,
        165, 166, 167, 168, 169,
        171, 172
Barany, I., 259, 261, 262
Baras, J. S., 114, 205, 208, 237,
        250, 265
Barrat, A., 117, 172, 173, 174,
        175, 176, 211
Barthélemy, M., 117, 172, 173,
        174, 175, 176, 211
Batagelj, V., 265
Bavelas, A., 124
Beauchamp, M. A., 126
Bertsekas, D. P., 75, 114, 225
Bertsimas, D., 79

Bianconi, G., 168, 169, 172
Birattari, M., 70
Blough, D. M., 191
Boguná, M., 142, 176, 177, 178,
        180, 185
Bollobas, B., 36, 45, 49
Bonabeau, E., 160
Bonacich, P., 135
Bonato, A., 167
Borgatti, S., 128
Borgatti, S. P., 124, 135
Boyd, S., 75
Brandes, U., 130, 131
Brin, S., 136
Broustis, I., 3

C

Calderbank, A. R., 19
Cavalcanti, D., 249
Cha, S., 58, 59
Chai-Keong, T., 3
Chang, J.-W., 201, 204
Chavel, I., 142
Chen, C., 201, 204
Chenyang, L., 197
Chiang, C. C., 3
Chiang, M., 19
Chiang, T. C., 201, 204
Chitradurga, R., 197, 198, 200,
        204
Cho, B. I., 208
Chung, K.-S., 203, 208, 222
Cormen, T. H., 114

D

Daly, E. M., 250

Deshpande, M., 128, 134
Diestel, R., 36
Dijkstra, E. W., 114
Dimokas, N., 250
Dodds, P. S., 149
Dorigo, M., 70
Dorogovtsev, S. N., 147, 168, 169
Dousse, O., 259
Doyle, J. C., 19
Durrett, R., 259

E
Easley, D., 115, 158
Ebeling, W., 86
Edling, C. R., 117
Eiben, A. E., 66, 87
Ephremides, A., 222
Eppstein, D., 127
Estrada, E., 182, 183, 184
Eun, D. Y., 250
Eunmi, C., 1

F
Fagiolo, G., 119, 121
Feller, W., 259
Fogel, D. B., 66, 74
Fogel, L. J., 98
Fortunato, S., 135, 136
Franceschetti, M., 259
Fredman, M. L., 128
Freeman, L., 128, 129
Freeman, L.C., 124, 128
Friedkin, N. E., 124, 126, 135
Friedl, B., 124, 126, 128, 135
Fronczak, A., 147
Fronczak, P., 147

G
Gallager, R. G., 114
Garcia-Molina, H., 237
Geisberger, R., 128, 132
Gerla, M., 3
Gkantsidis, C., 184
Goodman, D., 18
Gopalakrishnan, S., 26

Grimmett, G., 259
Grimmett, G. R., 45
Guidoni, D. L., 206, 208
Guimerà, R., 162, 185
Guoliang, X., 197
Gupta, P., 191
Gupta, V., 250

H
Haahr, M., 250
Habetler, T. G., 1
Haenggi, M., 259
Harary, F., 36
He, D., 1
Heidemann, J., 124, 126, 128, 135
Helmy, A., 197, 198, 200, 202,
          203, 204, 208, 223
Holland, J. H., 66, 79
Holme, P., 117
Holyst, J. A., 147
Hoory, S., 184, 185
Horvath, S., 118
Hovareshti, P., 205, 208, 250
Hu, B., 172
Hwang, Y. J., 208

J
Jackson, M. O., 231
Jakllari, G., 3
Jan, 201, R.-H., 204
Jeong, H., 156, 157, 161, 163, 172
Jiang, C.-J., 201, 204

K
Kamvar, S. D., 237
Karaliopoulos, M., 18
Karyotis, V., 191, 205, 206, 208,
          221, 222, 225, 226, 229,
          235
Kaski, K., 117, 118
Katsaros, D., 250
Kelner, J., 249
Kendall, W. S., 259
Kernighan, B., 82
Kertész, J., 117, 118

Kim, B. J., 117
Kim, S.-L., 208
Kintali, S., 127, 128, 131, 134
Kitsak, M., 176, 180, 185
Kivelä, M., 117, 118
Klain, D. A., 259, 261
Kleinberg, J., 115, 153, 154, 158
Ko, S.-W., 208
Krapivsky, P. L., 167, 168, 169
Krioukov, D., 142, 176, 177, 178, 180, 185
Kulkarni, R. V., 147
Kulkarni, S. R., 196, 197, 208
Kumar, P. R., 191
Kushafnagar, N., 250

L
Landherr, A., 124, 126, 128, 135
Lee, C.-H., 250
Lee, E. A., 26
Lee, S. I., 208
Leiserson, C. E., 114
Leoncini, M., 191
Leyvraz, F., 167, 168, 169
Lin, S., 82
Lin, W. S., 140
Lin, Y., 250
Linial, M., 184, 185
Liu, B., 208
Liu, K. J. R., 140
Liu, W., 3
Liu, Z., 208
Loureiro, A. A. F., 206, 208
Low, S. H., 19
Lui, S., 26
Lumb, I., 1
Lycan, W. G., 8

M
Madduri, L., 127, 128, 131, 134
Malliaros, F. D., 183
Manoj, B. S., 203, 204, 208
Matias, Y., 185
Mayer, C. P., 18

Mazumdar, R., 197, 198, 199, 200, 201
Mecke, J., 259
Megalooikonomou, V., 183
Mehrotta, S., 128, 134
Mendes, J. F. F., 147, 168, 169
Mickalewicz, Z., 74
Mihail, M., 127, 128, 131, 134, 184
Milgram, S., 148, 149
Mini, R. A. F., 206, 208
Mohri, M., 114, 265
Molle, M., 3
Morton, D., 1
Mossa, S., 162, 185
Muhamad, R., 149

N
Narayan, O., 141, 142
Neda, Z., 156, 157, 163
Nedic, A., 75
Newman, M., 5, 156
Nguyen, G. D., 222

O
Olfati-Saber, R., 250
Onnela, J.-P., 117, 118
Opsahl, T., 117, 124, 126, 128
Owens, A. J., 98
Ozdaglar, A. E., 75

P
Page, L., 136
Pantazopoulos, P., 18
Panzarasa, P., 117
Papadopoulos, F., 142, 176, 177, 178, 180, 185
Papavassiliou, S., 166, 191, 205, 206, 208, 221, 222, 225, 226, 229, 235
Papoulis, A., 158, 259
Park, S. M., 117
Pastor-Satorras, R., 117, 172
Penrose, M., 110
Perra, N., 135, 136
Pich, C., 131

Pillai, S. U., 58, 59, 158, 259
Pless, R., 197

Q
Qingfeng, H., 197
Qixin, W., 26

R
Rangarajan, A., 250
Rao, R., 203, 204, 208
Ravasz, E., 156, 157, 163
Rechenberg, I., 66, 86
Reddy, A., 157, 191
Redner, S., 163, 167, 168, 169
Rennie, J., 7
Repantis, T., 3
Resta, G., 191
Rezaei, B. A., 250
Reznik, A., 196, 197, 208
Rimal, B. P., 1
Riordan, O., 45
Rivest, R. L., 114
Rodríguez-Velázquez, J. A., 182, 183
Rota, G.-C., 259, 261
Rotman, J. J., 56
Roychowdhury, V. P., 250
Royer, E. M., 3
Rudolph, L., 1
Rui, X., 1

S
Saberi, A., 184
Sadok, D., 249
Samukhin, A. N., 168, 169
Sanders, P., 128, 132
Saniee, I., 141, 142
Santi, P., 191
Saramäki, J., 117, 118
Sarshar, N., 250
Schlosser, M. T., 237
Schneider, R., 259, 261, 262
Schubert, A., 156, 157, 163
Schultes, D., 128, 132
Schwefel, H.-P., 86

Schwefel, P.-H., 66
Serrano, 180, 185
Shakkottai, S., 157, 191
Sharma, G., 197, 198, 199, 200, 201
Singer, I. M., 142
Singh, H., 250
Singh, S., 250
Skvoretz, J., 124, 126, 128
Smith, J. E., 66, 87
Stai, E., 205, 206, 208, 221, 222, 225, 226, 229, 235
Stauffer, D., 45
Stavrakakis, I., 18
Stirzaker, D., 259
Stoyan, D., 259
Strogatz, S. H., 4, 150, 156, 157, 160, 196
Stroud, D., 147
Stutzle, T., 70
Suel, T., 58, 59

T
Tamma, B. R., 203, 204, 208
Tarjan, R. E., 128
Tassiulas, L., 250
Theodorakopoulos, G., 114, 237, 265
Thorpe, J. A., 142
Towsley, D., 208
Travers, J., 148, 149
Trubin, V. A., 113
Tse, D., 191
Tsitsiklis, J., 79
Tu, Y., 172
Turtsci, A., 162, 185

V
Vahdat, A., 142, 176, 177, 178, 185
Vandenberghe, L., 75
Venkatasubramanian, N., 128, 134
Verdú, S., 196, 197, 208
Verma, C. K., 203, 204, 208

Vespignani, A., 117, 172, 173, 174,
        175, 176, 211
Vicsek, T., 156, 157, 163
Viswanath, P., 191
Voigt, H.-M., 86

W
Waldhorst, O. P., 18
Walsh, M. J., 98
Wang, B-H., 172
Wang, J., 127
Wang, W.-X., 172
Wattenhofer, R., 191
Watts, D. J., 142, 149, 150, 153,
        156, 157, 160, 194, 196,
        205, 252
Weil, W., 259, 261, 262
Weixuan, L., 1
White, D. R., 128
Widgerson, A., 184, 185
Wieselthier, J. E., 222
Wilson, R. A., 56
Wu, H.-K., 3

X
Xie, Y.-B., 172
Xing, B., 128, 134
Xue, L., 26

Y
Yarvis, M., 250
Ying, L., 157, 191
Yook, S. H., 172

Z
Zhang, B., 118
Zhao, H. V., 140
Zhou, J., 166
Zhou, T., 172
Zollinger, A., 191
Zorpette, G., 7

Subject Index

A

Absolute evidence, 103
Adaptive parameter control, 103
Adjacency matrix, 33
Advertisement packets, 200
AI, *see* Artificial Intelligence
Algebraic Graph Theory, 55–59
    adjacency matrix and graph
        Laplacian, 56–59
    connected graph, 57
    empty graph, 58
    Google search engine, 59
    Lagrangian, 57, 58
    nonempty graph, 57
    Perron–Frobenius Theorem,
        58, 59
    theorem, 57
Algorithm parameters, 100
Alon–Boppana theorem, 182
Anytime property, 77
Arithmetic crossover, 85
ARPANET, 23
Artificial Intelligence (AI), 95
Average path length, 114–115
    centrality, 115
    hop-count, 114
    node pairs, identification of,
        114
    use, 115

B

Barabási–Albert algorithm, 165
Betweenness centrality, 128, 235
Bipartite graph, 36, 55

C

CAN, *see* Complex Network
    Analysis
Candidate solution, 73
Capacity, 4
Centrality, 122–138
    adaptive sampling
        betweenness vertex
        centrality
        approximation, 133
    applications, 124
    approximation of closeness
        centrality, 127
    betweenness centrality,
        128–130
    betweenness centrality
        approximation methods,
        130–135
    bisection based
        approximation approach,
        131
    bisection sampling, 133
    closeness (path) centrality,
        125–128
    degree centrality, 123–125
    Directed Acyclic Graph, 130
    eigenvector centrality,
        135–136
    example of centralities'
        computation, 136–138
    FACE algorithm, 134
    linear scaling, 132
    maximum betweenness
        centrality of vertex, 129
    metric, 128

network edge nodes, 125
pivots, 131
primary spanning tree, 135
ranking of nodes, 137, 138
secondary spanning tree, 135
Single Source Shortest Path,
127, 130
Chromosome, 73
Clique, 34
Closeness centrality, 125
Clustering coefficient, 115–122
adjacency matrix, 120
category cycle, 121
definition, 115–116
directed triangle, 119
extension to directed graphs,
118–122
extension to weighted graphs,
116–118
global clustering coefficient,
116
local clustering coefficient,
118
network-wide clustering
coefficient, 122
weighted local clustering
coefficient, 117
CNA, *see* Complex network
analysis
Co-evolution, 105
Cognitive methods and
evolutionary computing,
65–107
brief history of evolutionary
computing, 66
elements from evolution
theory, 66–68
genetic drift, 67
genome, 68
genotype, 67
natural selection, 66
phenotype, 67
population fitness, 66
trial-and-error approach,
67

evolutionary computing,
68–79
anytime property, 77
candidate solution, 73
chromosome, 73
components of evolutionary
algorithms, 70–72
exploitation, 76
exploration, 76
fitness function, 73
general mechanism of
evolutionary algorithm,
71
genotypes, 72
heuristics, 78
initialization and
termination conditions,
75–76
internal models, 68
mutation, 74
operation of evolutionary
algorithm, 76–79
optimization problem, 68
outputs, 68
parent selection, 74
population, 73–74
premature convergence, 76
pseudocode table, 72
representation, 72–73
simulation problem, 69
survivor selection, 75
system identification
problem, 69
trial-and-error, 71
unary operator, 74
variation operators
(recombination and
mutation), 74–75
evolutionary computing
approaches, 79–107
absolute evidence, 103
adaptive parameter
control, 103
algorithm parameters, 100
arithmetic crossover, 85

change-evidence plan, 104
co-evolution, 105–106
competitive fitness
    evaluation, 106
creep mutation, 82
cycle crossover, 87
data fitting problems, 98
deterministic parameter
    control, 102
discrete recombination, 85
edge crossover, 87
elitism, 91
evolutionary computing at
    a glance, 100
evolutionary programming,
    98–100
evolutionary strategy,
    91–95
external influence concept,
    105
feedback, 102
Finite State Machine, 98
fitness proportional
    selection, 89
"fixed" parameter control,
    103
fuzzy rule, 103
Gaussian distribution, 92
generational gap, 88
generational model, 88
genetic algorithms, 79–91
genetic programming,
    95–98
insert mutation, 82
integer representations, 80
interactive evolution,
    106–107
intermediate crossover, 85
inversion mutation, 82
multi-parent
    recombination, 86
mutation, 81, 93
mutualism, 105
next generation, 88
N-point crossover, 84

one-point crossover, 84
order crossover, 87
parameter control, 100–104
parameter timing, 100
parasitism, 105
parent selection, 89
parse trees, 95
partially mapped crossover,
    86
phenotype representation,
    107
population model, 88
positional bias, 85
predation, 105
random resetting, 82
ranking selection, 89
recombination, 83
relative evidence, 103
replacement worst, 91
scramble mutation, 82
self-adaptation, 91, 94, 103
"sigma last" strategy, 99
simple recombination, 85
simulated annealing, 79
single arithmetic
    recombination, 86
special forms of evolution,
    105–107
steady-state model, 88
"stochastic search" degree,
    106
stochastic universal
    sampling, 90
survivor selection, 90
swap mutation, 82
symbiosis, 105
tournament selection, 90
uniform arithmetic
    recombination, 85
uniform crossover, 84
uniform mutation, 82
whole arithmetic
    recombination, 86
evolutionary cycle, 65
genetic programming, 66

Coloring, 53–55
    bipartite graph, 55
    chromatic number, 53
    description of coloring, 53
    edge-chromatic number, 53
    independent sets, 53
    plane graph, 54
    theorem, 53, 54
Commute time, 41
Competitive fitness evaluation,
    106
Complex communication
    networks, conclusion,
    247–256
    epilogue, 255
    lessons learned, 247–251
        discussion on evolutionary
        topology modification
        mechanisms, 251
        emerging trends and their
        benefits, 248–251
        evolutionary network
        modification framework,
        249
        network lifetime, 250
        Network Science, research,
        248
        scale-free (power-law)
        regime, 249
        small-world paradigm, 248
        socially-inspired topology
        modification
        mechanisms, 250
        "stochastic search"
        approach, 249
    road ahead, 251–254
        centrality measures, 252
        invention of centrality
        definition, 254
        inverse SETM, 253
        network performance, 253
        node strength distribution,
        251
        open problems, 253–254
        QoS, 253
        route covered already,
        251–253
        SETM, 254
        structure of social
        networks, 252
        topology modification
        methodology, 253
Complex communication
    networks, introduction,
    1–27
    approach and objectives, 2–4
        application frameworks, 4
        goal-topics, 4
        hierarchical approach, 3, 4
        latest trends, 3
        stringent environments, 3
    fundamentals of complex
        networks, 5–18
        balance of trade-offs, 15
        characterization of complex
        networks, 14
        completely engineered
        complex networks, 15
        complex network analysis,
        18
        complex network
        fundamentals, 10–12
        complex network taxonomy
        and examples, 12–18
        definition of complex
        networks, 5
        definition of network
        formation, 10
        dominating features of
        complex networks, 6
        efficiency/convenience, 16
        human-initiated complex
        networks, 15
        interaction, 6
        MAC layer protocol, 7
        modeling of network types,
        9
        modern societies, 5
        natural networks, 14
        nonpredictable cumulative

behaviors, 7
origin criterion, 13
performance indices, 13
power-law, 16
Quality of Service, 8
trade-off between benefit
vs. cost of collaboration,
11
Network Science, 18–27
ARPANET, 23
common core, 22
content and promise of
network science, 21–22
cross-disciplinary research,
24
DARPA, 23
definition of Network
Science, 19
evolution, 20
first networking structures,
23
fragmentation, 20
holistic perspective, 19
human understanding of
networks, 24
initial question, 21
major classes of challenges,
26
mathematical
methodologies, 24–25
networks and network
research in the 21st
century, 23–25
Network Utility
Maximization, 19
new application domains,
25
online social networks, 21
status and challenges of
network science, 25–27
"umbrella-like" areas, 26
Complex Network Analysis
(CNA), 18, 109
Complex networks, 1
Complex network taxonomies, 14

Complex and social network
analysis metrics and
features, 109–144
average path length, 114–115
centrality, 115
hop-count, 114
node pairs, identification
of, 114
use, 115
centrality, 122–138
adaptive sampling
betweenness vertex
centrality
approximation, 133
applications, 124
approximation of closeness
centrality, 127
betweenness centrality,
128–130
betweenness centrality
approximation methods,
130–135
bisection based
approximation approach,
131
bisection sampling, 133
closeness (path) centrality,
125–128
degree centrality, 123–125
Directed Acyclic Graph,
130
eigenvector centrality,
135–136
example of centralities'
computation, 136–138
FACE algorithm, 134
linear scaling, 132
maximum betweenness
centrality of vertex, 129
metric, 128
network edge nodes, 125
PageRank, 136
pivots, 131
primary spanning tree, 135
ranking of nodes, 137, 138

secondary spanning tree, 135
Single Source Shortest Path, 127, 130
clustering coefficient, 115–122
  adjacency matrix, 120
  category cycle, 121
  definition, 115–116
  directed triangle, 119
  extension to directed graphs, 118–122
  extension to weighted graphs, 116–118
  global clustering coefficient, 116
  local clustering coefficient, 118
  network-wide clustering coefficient, 122
  weighted local clustering coefficient, 117
Complex Network Analysis, 109
curvature, 141–143
  definition, 141
  Gauss–Bonnet theorem, 142
  global network feature, 141
  hyperbolicity, 141
  models, 142
  negative curvature, 141
  planar graphs, 142
degree distribution, 110–112
  degree sequence, 112
  forms, 110
  incorrect reference, 111
  neighborhood relationships, 110
  network topology, 112
metrics at a glance, 143–144
paradigms, 110
prestige, 138–140
  degree prestige, 139
  influence domain, 139–140
  intuitive definition, 138

  proximity prestige, 140
  sink vertex, 139
  Social Network Analysis, 109
  strength, 113
    in-strength, 113
    out-strength, 113
    vertex strength, 113
Connectivity, 38–46
  commute, time, 41
  connectivity pair, 41
  cut-set, 44
  edge-disjoint paths, 42
  graph components, 42
  k-connectivity, 41
  line-connectivity, 39
  nonseparable graph, 45
  nontrivially connected graph, 46
  sparseness of topology, 40
  theorem, 40, 43
  trivial graph, 39
  Tutte's theorem, 41
Continuum theory, 166
Covering, 53–55
  bipartite graph, 55
  chromatic number, 53
  description of coloring, 53
  edge-chromatic number, 53
  independent sets, 53
  plane graph, 54
  theorem, 53, 54
Creep mutation, 82
Crossover, 83
Cut-set, 44
Cycle, 47
Cycle crossover, 87

D
DAG, *see* Directed Acyclic Graph
DARPA, 23
Data mules, 205
Degree centrality, 123–125
Degree distribution, 110–112
  degree sequence, 112
  forms, 110

incorrect reference, 111
neighborhood relationships, 110
network topology, 112
Delay, 190
Deterministic parameter control, 102
Diameter, 47
Directed Acyclic Graph (DAG), 130
Discrete recombination, 85
Distinctive structure and features of complex networks, 145–186
expansion properties, 180–185
Alon–Boppana theorem, 182
applications of expander graphs, 184–185
definition and analytical properties, 180–184
expansion ratio, 181
Petersen graph, 181
social networks, 183
hyperbolic structure of complex networks, 176–180
background on hyperbolic geometry, 176–177
Euclidean circle, 177
evolutionary models developed on hyperbolic geometry, 178–180
global shortest paths, 180
Poincaré Disk, 177
network structure and evolution, 145–146
application, 145
connectivity features, 146
node properties, 146
opening potentials, 145
scale-free networks, 158–176
Barabási–Albert algorithm, 165
Barabási–Albert model, 162–167
citation network, 161
continuum theory, 166
definition and properties, 158–160
Edge Churn, 169
events formalism, 170
examples and applications, 161–162
extensions of the Barabási–Albert model, 168–176
growth, 159, 165
holistic modification framework, 169
homogeneous coupling, 174
initial attractiveness, 168
Node Churn, 169
Poisson distribution, 159
power-law distributions, 159
preferential attachment, 159, 165
rate equation, 167
signature, 159
weighted and directed network graphs, 172
World Wide Web Graph, 175
zero-degree attractiveness, 168
small-world paradigm, 146–157
clustering coefficient, 147
examples and applications, 155–157
Internet-based social search experiment, 149
Kleinberg's model, 153–155
large-scale experiments, 148–150
Prolegomena (description of a small-world network), 146–148
relational graphs, 146

six degrees of separation,
148–150
small-world graphs, 151,
153
Watts and Strogatz model,
150–153
Dual function, 225

E
Edge addition process, 210
Edge crossover, 87
Edge deletion process, 213
EDS, *see* Energy Dissipation Skew
Eigenvector centrality, 135–136
Elitism, 91
Energy Dissipation Skew (EDS),
199
Eulerian circuit, 47
Eulerian trail, 47
Euler Polyhedron Formula, 51
Evolutionary algorithms, 66, 70
Evolutionary approaches, 187–245
Holistic Topology
Modification Framework,
208–228
assumption, 211
combined mechanism
(WEC and WNC), 220
Continuum Theoretic
approach, 212
deployment area, 209
determination of equations
for constraint set, 223
determination of objective
function, 222
dual function, 225
energy consumption, 228
link addition and deletion,
225
optimization formulation,
224
optimization methodology,
221–228
process of edge addition,
210

process of edge deletion,
213
process of node addition,
216
process of node deletion,
218
Protocol Stack, 213
Topology Modification
Mechanism, 209, 220
Weighted Edge Churn
framework, 209–216
Weighted Node Churn
framework, 216–220
inverse topology
control-based
approaches, 195–208
advantages and
disadvantages of using
wired shortcuts, 201
Advertisement packets, 200
cells, 198
data mules, 205
deterministic
shortcut-topology, 203
Energy Dissipation Skew,
199
energy efficiency, 197, 201
Euclidean distance, 203
greedy forwarding routing,
205
HELLO messages, 201
Quality of Service
requirements, 195
Random Geometric Graph,
202
Watts and Strogatz
mechanism, 201
wired link placement, 196
wired shortcuts,
approaches using,
196–202
wireless shortcuts,
approaches using,
202–208
spatial graphs and

small-world
phenomenon, 192–195
clustering coefficient, 194
links, 193
naive mechanism, 194
relational graphs, 193
shortcuts, 195
special cases, 229–244
betweenness centrality, 235
clustering coefficient, 232
elimination to binary
graphs (SETM), 229–235
global algorithm, 240
infusion of small-world
properties, 231
in-strength, 238
inverse preferential
attachment, 229
local algorithm, 239
shortest path
communication, 238
stochastic dominance, 231
topology modification
mechanism, 238
trust management in
wireless multi-hop
networks, 236–244
trust values, 236
topology control and inverse
topology control,
191–192
lack of optimization, 192
multi-hop networks, 191
QoS-relevant performance
metrics, 192
trade-off, 191
wireless multi-hop
communications,
188–191
assumption, 189
delay of message delivery,
190
Physical Model, 190
Random Geometric Graph,
189

self-organization, 188
Signal-to-Interference-plus-
Noise-Ratio,
190
throughput, 190
transport capacity, 191
unicast link, 189
Evolutionary computing, 68–79,
see also Cognitive
methods and
evolutionary computing
anytime property, 77
candidate solution, 73
chromosome, 73
components of evolutionary
algorithms, 70–72
exploitation, 76
exploration, 76
fitness function, 73
general mechanism of
evolutionary algorithm,
71
genotypes, 72
heuristics, 78
initialization and termination
conditions, 75–76
internal models, 68
mutation, 74
operation of evolutionary
algorithm, 76–79
optimization problem, 68
outputs, 68
parent selection, 74
population, 73–74
premature convergence, 76
pseudocode table, 72
representation, 72–73
simulation problem, 69
survivor selection, 75
system identification
problem, 69
trial-and-error, 71
unary operator, 74
variation operators
(recombination and

mutation), 74–75
Evolutionary programming,
        98–100
External influence concept, 105

F
FACE algorithm, 134
Features of complex networks, *see*
        Distinctive structure and
        features of complex
        networks
Finite State Machine (FSM), 98
Fitness, 66
Fitness function, 73
Fitness proportional selection, 89
"Fixed" parameter control, 103
Flow, 48–50
        capacity function, 50
        definition, 48
        finite directed graph, 48
        Kirchhoff's law, 49
        Max-Flow Min-Cut theorem,
                49
FSM, *see* Finite State Machine
Fuzzy rule, 103

G
GA, *see* Genetic Algorithms
Gain vs. cost of collaboration
        tradeoff, 10
Gauss–Bonnet theorem, 142
Generational gap, 88
Generational model, 88
Genetic Algorithms (GA), 79–91,
        95
Genetic drift, 67
Genetic programming, 66, 95
Genome, 68
Genotype, 67, 72
Geodesic path, 47
Geometric probability, 259–262
        definitions, 260, 261
        independent random
                variables, 261
        joint density function, 262

joint probability distribution
        function, 261
Markov property, 262
memoryless property, 262
probabilistic modeling of
        deployment of wireless
        multi-hop network,
        261–262
probability density function,
        261
probability distribution
        function, 260
probability space, 259
probability theory elements,
        259–261
Random Geometric Graphs,
        261
random variable, 260
spatial Markov property, 262
Girth, 47
Google, 59, 136
Graph models and their properties
        (basic network), 29–64
        additional definitions, 35–38
        bipartite graph, 36
        complete bipartite graph,
                36
        complete set of invaraints,
                36
        greedy algorithms, 38
        invariant, 36
        isomorphic graphs, 35
        Kruskal's and Prim's
                algorithm, 38
        spanning tree, 37, 38
        statements, 37
        Algebraic Graph Theory,
                55–59
        adjacency matrix and
                graph Laplacian, 56–59
        connected graph, 57
        empty graph, 58
        Google search engine, 59
        Lagrangian, 57, 58
        nonempty graph, 57

Perron–Frobenius
  Theorem, 58, 59
 theorem, 57
basic definitions and
    notation, 31–35
 adjacency matrix, 33, 35
 clique, 34
 complete graph, 34
 d-regular graph, 34
 incidence matrix, 34
 induced subgraph, 32
 maximal clique, 34
 spanning subgraph, 32
 subgraph, 32
 undirected graph, 31
coloring (covering), 53–55
 bipartite graph, 55
 chromatic number, 53
 description of coloring, 53
 edge-chromatic number, 53
 independent sets, 53
 plane graph, 54
 theorem, 53, 54
connectivity, 38–46
 commute, time, 41
 connectivity pair, 41
 cut-set, 44
 edge-disjoint paths, 42
 graph components, 42
 k-connectivity, 41
 line-connectivity, 39
 nonseparable graph, 45
 nontrivially connected
    graph, 46
 sparseness of topology, 40
 theorem, 40, 43
 trivial graph, 39
 Tutte's theorem, 41
flow, 48–50
 capacity function, 50
 definition of flow, 48
 finite directed graph, 48
 Kirchhoff's law, 49
 Max-Flow Min-Cut
    theorem, 49

graph theory fundamentals,
    30–59
 additional definitions,
    35–38
 Algebraic Graph Theory,
    55–59
 basic definitions and
    notation, 31–35
 coloring (covering), 53–55
 connectivity, 38–46
 description of graph, 31
 flow, 48–50
 paths and cycles, 46–48
 planarity, 50–52
 value of means provided, 30
notation, 63, 64
paths and cycles, 46–48
 ability to traverse, 46
 closed walk cycle, 47
 Eulerian circuit, 47
 Eulerian trail, 47
 geodesic path, 47
 girth, 47
 Hamilton cycle, 48
 Hamilton graph, 48
 walk of a graph, 46
planarity, 50–52
 embedded graph, 50
 Euler Polyhedron Formula,
    51
 exterior face, 51
 Kuratowski's theorem, 52
 planar graph, 51
 plane map, 51
random graphs, 59–63
 basic random graph
    models, 60–63
 fixed graph, 60
 Hamilton cycles, 62
 intuitive way of
    considering, 59
 monotone decreasing
    graph, 61
 monotone increasing graph,
    61

probabilistic methods, 59
    theorem, 61
    threshold functions, 62
Greedy forwarding routing, 205
Growth, 159

H
Hamilton cycle, 48, 62
Hamilton graph, 48
HELLO messages, 201
Holistic Topology Modification
        Framework, 208–228
    assumption, 211
    combined mechanism (WEC
        and WNC), 220
    Continuum Theoretic
        approach, 212
    deployment area, 209
    determination of equations
        for constraint set, 223
    determination of objective
        function, 222
    dual function, 225
    energy consumption, 228
    link addition and deletion,
        225
    optimization formulation, 224
    optimization methodology,
        221–228
    process of edge addition, 210
    process of edge deletion, 213
    process of node addition, 216
    process of node deletion, 218
    Protocol Stack, 213
    Topology Modification
        Mechanism, 209, 220
    Weighted Edge Churn
        framework, 209–216
    Weighted Node Churn
        framework, 216–220
Hop-count, 114
Hyperbolic geometry, 176–177
Hyperbolicity, 141

I
Incidence matrix, 34
Independent random variables,
        261
Independent sets, 53
Initial attractiveness, 168
Insert mutation, 82
Interactive evolution, 106
Intermediate crossover, 85
Inverse SETM, 253
Inverse Topology Control (iTC),
        191
Inverse Topology Control-based
        approaches, 195–208
    advantages and disadvantages
        of using wired shortcuts,
        201
    Advertisement packets, 200
    cells, 198
    data mules, 205
    deterministic
        shortcut-topology, 203
    Energy Dissipation Skew, 199
    energy efficiency, 197, 201
    Euclidean distance, 203
    greedy forwarding routing,
        205
    HELLO messages, 201
    Quality of Service
        requirements, 195
    Random Geometric Graph,
        202
    Watts and Strogatz
        mechanism, 201
    wired link placement, 196
    wired shortcuts, approaches
        using, 196–202
    wireless shortcuts, approaches
        using, 202–208
Inversion mutation, 82
iTC, *see* Inverse Topology Control

J
Joint probability distribution
        function, 261

K
Kirchhoff's law, 49
Kleinberg model, 153–155

L
Lagrangian, 57, 58
Laplacian, 56–59
LISP programming language, 95

M
MAC layer protocol, 7
Markov property, 262
Max-Flow Min-Cut theorem, 49
Memoryless property, 262
Monoid, 263–264
Multi-parent recombination, 86
Mutation, 74, 81
Mutualism, 105

N
Natural selection, 66
Negative curvature, 141
Network evolutionary dynamics, 4
Network graph models, *see* Graph
    models and their
    properties (basic
    network)
Network Science, 1–2, 18–27
    ARPANET, 23
    common core, 22
    content and promise of
        network science, 21–22
    cross-disciplinary research, 24
    DARPA, 23
    definition of Network Science,
        19
    evolution, 20
    first networking structures, 23
    fragmentation, 20
    holistic perspective, 19
    human understanding of
        networks, 24
    initial question, 21
    major classes of challenges,
        26
    mathematical methodologies,
        24–25
    networks and network
        research in the 21st
        century, 23–25
    Network Utility
        Maximization, 19
    new application domains, 25
    online social networks, 21
    status and challenges of
        network science, 25–27
    "umbrella-like" areas, 26
Network Utility Maximization
    (NUM), 19
Next generation, 88
Node addition, 216
Node deletion, 218
Node pair distance, 47
N-point crossover, 84
NUM, *see* Network Utility
    Maximization

O
One-point crossover, 84
Optimization methodology,
    221–228
Order crossover, 87
Origin criterion, 13

P
PageRank, 136
Parameter control, 100
Parasitism, 105
Parent selection, 74, 89
Parse trees, 95
Partially mapped crossover, 86
Path, 47
Perron–Frobenius Theorem, 58, 59
Petersen graph, 181
Phenotype, 67
Physical Model, 190
Planar graph, 51, 142
Poincaré Disk, 177
Population, 73–74
Population fitness, 66

Population model, 88
Positional bias, 85
Predation, 105
Preferential attachment, 159
Prestige, 138–140
    degree prestige, 139
    influence domain, 139–140
    intuitive definition, 138
    proximity prestige, 140
    sink vertex, 139
Primary spanning tree (PST), 135
Probability density function, 261
Probability distribution function,
    260
PST, *see* Primary spanning tree

Q
QoS, *see* Quality of Service
Quality of Service (QoS), 8, 195,
    202

R
Random Geometric Graph
    (RGG), 189, 202, 261
Random graphs, 59–63
    basic random graph models,
        60–63
    fixed graph, 60
    Hamilton cycles, 62
    intuitive way of considering,
        59
    monotone decreasing graph,
        61
    monotone increasing graph,
        61
    probabilistic methods, 59
    theorem, 61
    threshold functions, 62
Random variable, 260
Ranking selection, 89
Recombination, 74, 83
Regular graph, 34
Relational graphs, 146, 193
Relative evidence, 103
Replacement worst, 91

Representation, 72–73
RGG, *see* Random Geometric
    Graph

S
SA, *see* Simulated annealing
Scale-free networks, 158–176
    Barabási–Albert algorithm,
        165
    Barabási–Albert model,
        162–167
    citation network, 161
    continuum theory, 166
    definition and properties,
        158–160
    Edge Churn, 169
    events formalism, 170
    examples and applications,
        161–162
    extensions of the
        Barabási–Albert model,
        168–176
    growth, 159, 165
    holistic modification
        framework, 169
    homogeneous coupling, 174
    initial attractiveness, 168
    Node Churn, 169
    Poisson distribution, 159
    power-law distributions, 159
    preferential attachment, 165
    rate equation, 167
    signature, 159
    weighted and directed
        network graphs, 172
    World Wide Web Graph, 175
    zero-degree attractiveness,
        168
Scramble mutation, 82
Secondary spanning tree (SST),
    135
Self-adaptation, 91,94
Self-adaptive parameter control,
    103
Semirings and path problems,

263–265
adaptation of semiring, 263
algebraic structure, 264
examples, 265
generalized shortest path, 264
monoids, 263–264
semirings, 264
traffic between nodes, 265
trust values, 265
Shortest path communication, 238
"Sigma last" strategy, 99
Signal-to-Interference-plus-Noise-
Ratio (SINR),
190
Simple recombination, 85
Simulated annealing (SA), 79
Simulation problem, 69
Single arithmetic recombination,
86
Single Source Shortest Path
(SSSP), 127, 130
SINR, *see* Signal-to-Interference-
plus-Noise-Ratio
Six degrees of separation, 148–150
Small-world, 150
Small-world paradigm, 146–157
clustering coefficient, 147
examples and applications,
155–157
Internet-based social search
experiment, 149
Kleinberg's model, 153–155
large-scale experiments,
148–150
Prolegomena (description of
a small-world network),
146–148
relational graphs, 146
six degrees of separation,
148–150
small-world graphs, 151, 153
Watts and Strogatz model,
150–153
SNA, *see* Social Network Analysis
Social Network Analysis (SNA),

109, *see also* Complex
and social network
analysis metrics and
features
Spanning tree, 37, 38
Spatial graphs, 192–195
Spatial Markov property, 262
SSSP, *see* Single Source Shortest
Path
SST, *see* Secondary spanning tree
Steady-state model, 88
Stochastic dominance, 231
Stochastic universal sampling, 90
Strength, 113
Subgraph, 32
Survival, 66
Survivor selection, 75
Swap mutation, 82
Symbiosis, 105
System identification problem, 69

T
Throughput, 190
Topology Control (TC), 191, *see
also* Inverse Topology
Control-based
approaches
Topology Modification
Mechanism, 209, 220,
238
Tournament selection, 90
Transport capacity, 191
Trust values, 236, 265
Tutte's theorem, 41

U
Unicast link, 189
Uniform arithmetic
recombination, 85
Uniform crossover, 84
Uniform mutation, 82

V
Vertex strength, 113

W

Walk, 46
Watts and Strogatz (WS) model,
        150–153
WEC, *see* Weighted Edge Churn
Weighted Edge Churn (WEC),
        208, 209
Weighted Node Churn (WNC),
        208, 216
Whole arithmetic recombination,
        86
Wired shortcuts, 196
Wireless multi-hop
        communications,
        188–191
    assumption, 189
    delay of message delivery, 190
    Physical Model, 190
    Random Geometric Graph,
        189
    self-organization, 188
    Signal-to-Interference-plus-
        Noise-Ratio,
        190
    throughput, 190
    transport capacity, 191
    unicast link, 189
Wireless shortcuts, 202
WNC, *see* Weighted Node Churn
WS model, *see* Watts and
        Strogatz model, 150–153

Z

Zero-degree attractiveness, 168